Representation and Reality

Representation and Mind
Hilary Putnam and Ned Block, editors

Representation and Reality, *by Hilary Putnam*
Explaining Behavior: Reasons in a World of Causes, *by Fred Dretske*

Representation and Reality

Hilary Putnam

A Bradford Book
The MIT Press
Cambridge, Massachusetts
London, England

This book was set in Palatino
by Graphic Composition, Inc., Athens, Georgia
and printed and bound by Halliday Lithograph
in the United States of America.

Library of Congress Cataloging-in-Publication Data

Putnam, Hilary.
  Representation and reality.
  "A Bradford book."
  Bibliography: p.
  Includes index.
  1. Realism. 2. Mind-brain identity theory. 3. Functionalism (Psychology). 4. Truth. 5. Reference (Philosophy). 6. Mathematics—Philosophy. 7. Computers. I. Title.
  B835.P88   1988   128'.2   87-35279
  ISBN 0-262-16108-7

For Burton Dreben,
who still won't be satisfied

Contents

Preface

This book is primarily a criticism of currently fashionable philosophical views held in and around the cognitive science community. They are the views of philosophers, including some of my former selves, but they are by no means held only by philosophers. I am dissatisfied with these views, and so this book consists of philosophical criticism, but I am by no means depressed by what some will regard as the "negative" outcome of my investigations. As I suggest in the last chapter, it is only by seeing that the currently fashionable views do not work that we can begin to see what the tasks of philosophy might really be.

I was enabled to start work on what became this book by the generosity of the National Endowment for the Humanities, which gave me a fellowship in 1982–1983. I was able to try out various versions of the book in different lecture series that I was invited to give. One of the earliest versions was tried out in Princeton, where I was briefly a Visiting Senior Fellow in the Humanities (in 1985) and had valuable opportunities to discuss my criticism of functionalism with Gil Harman, Saul Kripke, and David Lewis, among others. Later I tried out other versions at Tel Aviv University and at the University of Munich, where I received many valuable comments and criticisms. The final version of three chapters (chapters 1, 5, and 7) formed the substance of my Whidden Lectures at McMaster University in the fall of 1987. All of my colleagues have in one way or another contributed to my thinking on this topic (and none can be held responsible for the results). In particular, Burton Dreben persuaded me to undertake a radical reworking of the penultimate version. An earlier version was substantially rewritten as the result of criticisms by two close readers: Charles Travis and my dearest critic Ruth Anna Putnam. And I owe many thanks to the participants in my 1986 NEH Summer Seminar.

Introduction

Many years ago I was invited to give a lecture on what is today called "computer science" at a large eastern university. I titled my lecture "Turing Machines," because the most famous abstract model of a computer is the model produced by Alan Turing. Today biographies of Turing are reviewed in the *New York Times*, but in those early days of the computer Turing was virtually unheard of. Thus it wasn't surprising that someone at the university "corrected" what he assumed to be my typographical error, with the result that posters announcing that I would give a lecture on TOURING MACHINES went up all over the campus. (A few people left rather early in the lecture.)

My interest in computers and the mind thus dates from a very early period. I may have been the first philosopher to advance the thesis that the computer is the right model for the mind. I gave my form of this doctrine the name "functionalism," and under this name it has become the dominant view—some say the orthodoxy—in contemporary philosophy of mind.

In this book I shall be arguing that the computer analogy, call it the "computational view of the mind," or "functionalism," or what you will, does not after all answer the question we philosophers (along with many cognitive scientists) want to answer, the question "What is the nature of mental states?" I am, thus, as I have done on more than one occasion, criticizing a view I myself earlier advanced. Strangely enough, there are philosophers who criticize me for doing this. The fact that I change my mind in philosophy has been viewed as a character defect. When I am lighthearted, I retort that it might be that I change my mind so often because I make mistakes, and that other philosophers don't change *their* minds because *they* simply never make mistakes. But I should like now, for once, to say something serious about this. I have never forgotten the conversations I had with Rudolf Carnap in the years 1953–1955, and in particular, I have never forgotten how Carnap—a great philosopher who had an aura of integrity and seriousness which was almost overwhelming— would stress that he had changed his mind on philosophical issues, and changed it more than once. "I used to think . . . I *now* think" was a sentence construction that was ever on Carnap's lips. And, of course, Russell, who influenced Carnap as Carnap influenced me, was also criticized for changing his mind. Although I do not now agree with Carnap's doctrines of any particular period, for me Carnap is still the outstanding example of a human being who puts the search for truth higher than personal vanity. A philosopher's job is not to produce a view X and then, if possible, to become universally known as "Mr. View X" or "Ms. View X." If philosophical investigations (a phrase made famous by another philosopher who "changed his mind") contribute to the thousands-of-years-old dialogue which is philosophy, if they deepen our understanding of the riddles we refer to as "philosophical problems," then the philosopher who conducts those investigations is doing the job right. Philosophy is not a subject that eventuates in final solutions, and the discovery that the latest view—no matter if one produced it oneself—*still* does not clear away the mystery is characteristic of the work, when the work is well done. I could add that what I just described as "changing my mind" is not a matter of "conversion" from one view to another; it is rather a matter of being torn between opposing views of the nature of philosophy itself. When I was a "scientific realist," I felt deeply troubled by the difficulties with scientific realism; having given up scientific realism, I am still tremendously aware of what is appealing about the scientific realist conception of philosophy. I hope that the present book at least partly reveals this "being torn."

But enough of this. The computational view was itself a reaction against the idea that our matter is more important than our function, that our *what* is more important than our *how.* My "functionalism" insisted that, in principle, a machine (say, one of Isaac Asimov's robots), a human being, a creature with a silicon chemistry, and a disembodied spirit could all work much the same way when described at the relevant level of abstraction, and that it is just wrong to think that the essence of our minds is our "hardware." This much—and it was central to my former view—I shall not be giving up in this book, and indeed it still seems to me to be as true and as important as it ever did. What I shall try to do is the trick attributed to adepts in jujitsu of turning an opponent's strength against himself: I shall try to show that the arguments for the computational view, in fact, the very arguments I formerly used to show that a simpleminded identification of mental states with physical-chemical states cannot be right, can be generalized and extended to show that a straightforward identification of mental states with *functional* states, i.e., with computationally characterized states, also cannot be right. Functionalism argued that mental states cannot simply *be* physical-chemical states, although they are emergent from and supervenient on physi-cal-chemical states; I shall now argue that mental states also cannot be computational states, or computational cum physical states (states defined using a mixed vocabulary referring both to physical and to computational parameters), although they are emergent from and may be supervenient upon our computational states. Unlike my *Reason, Truth and History*, this book does not suggest a general stance toward metaphysical questions, and, apart from a brief sketch in the final chapter, I shall try to keep my own "positive" views out of the work. Although there are many parts to the argument I will be pre-senting, it was developed as a single argument. In the future I hope to return to larger metaphysical questions; here I aim at giving a fairly complete account of one particular line of thought on one particular philosophical issue, digressing (as I shall have to) into issues in the philosophy of language, the theory of causation, the nature of truth, and so on, only to the extent that those issues bear on the chosen topic.

The first three chapters actually grew out of two earlier papers.[1] Those papers were, in part, polemics against the views of my good friend and former student Jerry Fodor. Fodor, I hasten to say, is *not* the main target of this book; but I have retained some of my polemic against what I call "MIT mentalism," because the arguments are drawn on in later chapters. The main target of the present book is one H. Putnam (one of my former selves) and those who have adopted his views. Or perhaps it would be more accurate to say that the present book doesn't have a "main target"; for its aim is not so much to refute one particular view as to establish the need for a dif-ferent way of looking at problems about "mental states." At any rate, the intended contribution of these three chapters to that end is to do two things: (1) to establish a close connection (discovered and em-phasized throughout his career by W. V. Quine) between problems about meaning and problems about belief fixation, by showing that the holistic character of belief fixation in science bears deeply on the issue of the individuation of "meanings" (or "contents" or "inten-tions," as they are called by various philosophers); and (2) to argue that, in fact, thinking of "meanings" (or "contents") as "theoretical entities"—as scientific objects, objects which can be isolated and which can play an explanatory role in a scientific theory—is a mis-take. In the course of the argument I defend the view that there is no criterion for sameness of meaning except actual interpretative prac-tice—a view made famous by Quine and Davidson.

Chapter 4 was difficult to place, because it is really on a "parallel track" to the rest of the volume; yet I found it impossible to omit. One influential line of thought in recent years maintains that what the difficulties with individuating (/giving a scientific account of) either propositional attitudes or "meanings" show is that talk of both belongs to "folk psychology." While some philosophers take folk psychology seriously as an explanatory theory, the "eliminationist" philosophers (e.g., Quine) are prepared to dismiss it as "second-class" talk, useful, perhaps, when we are doing "personal biography" but having no place in the description of Nature (which alone has metaphysical import, according to these philosophers).

The best response to such an argument is to point out that the difficulties with functionalist views (developed in chapters 5 and 6) apply as much to "physicalist" accounts of reference as to "physicalist" accounts of meaning. Reference is the main tool used in formal theories of truth. But *truth* is not just a notion of folk psychology; it is the central notion of *logic*. None of these philosophers wishes to give up logic. Eliminationist philosophers must meet this challenge—the challenge of showing that their "let's *eliminate* talk of the mental from our metaphysical picture" stance doesn't require the "elimination" of the notion of truth. Generally they try to do this (*a*) by saying that Tarski showed that the notion of "truth" can be defined without appealing to any dubious mentalistic or "intentional" notions; or (*b*) by claiming that truth is just a device for "disquotation."

If I did not respond to these views, then, I knew, my entire book would evoke a "We told you so" from the eliminationists. Hence the need for a chapter devoted to questions about truth. If I am right, the idea that there can be an account of truth which has "nothing to do with the mental" is an illusion.

Chapters 5 and 6 build on the previous material, especially on the arguments for meaning holism. The purpose of these chapters is to argue that mental states are not only compositionally plastic (the same "mental state" can, in principle, be a property of systems which are not of the same physical constitution) but *computationally* plastic as well—the same mental state (e.g., the same belief or desire) can in principle be a property of systems which are not of the same computational structure. Mental states cannot literally be "programs," because physically possible systems may be in the same mental state while having unlike "programs."

This leads to the difficult question whether there is nevertheless a kind of "equivalence" between the structures of all physically possible systems (organisms cum environments) which contain a physically possible organism who entertains a particular belief; a kind of equivalence which can be defined in physical cum computational terms. These chapters present an argument designed to show that if such an equivalence relation existed, it would be undiscoverable—not just undiscoverable by human beings, but *undiscoverable by physically possible intelligent beings.*

One may want to say that, even if this is right, such a relation might still exist. It might be, someone might claim, that *believing there are a lot of cats in the neighborhood* (substitute your favorite example of a belief here) is just being part of a "system"—an environment plus society of organisms—which may be in any one of a number of computational cum physical conditions, conditions which are, however, "equivalent" in the sense defined by this (unknown and, if I am right, *unknowable*) relation.

This is exactly analogous to saying that the true nature of *rationality*—or at least of human rationality—is given by some "functional organization," or computational description. We know from arguments of the kind made famous by Kurt Gödel[2] that if there is such a description, then we could never justify the claim that the description is correct (by methods of justification formalized by the description itself). But if the description is a formalization of our powers to reason rationally *in toto*—a description of *all* our means of reasoning—then inability to know something by the "methods formalized by the description" is inability to know that something *in principle.* Here too some philosopher might say, "Even so there is such a description. It doesn't matter that we can't tell which one it is."

The difficulty with this claim, and with all such claims, is not that physically possible organisms don't have functional organizations, but that they have *too many.* A theorem proved in the Appendix to this book shows that there is a sense in which *everything has every functional organization.* When we are correctly described by an infinity of logically possible "functional descriptions," what is the claim supposed to *mean* that one of these has the (unrecognizable) property of being our "normative" description? Is it supposed to describe, in some way, our very *essence?*

This question may be a natural one with which to close the present work. If we see that attempts to "naturalize" reference (in the sense of showing that it is just another "physical relation") lead back to just the metaphysical obscurities that such accounts were designed to clear up, then we can see the need for a different—and better—way of thinking about these philosophical issues.

Representation and Reality

Chapter 1
Meaning and Mentalism

Today someone reading a review of a philosophical book in the *New York Review of Books* might well encounter the word "intentionality." Yet few reviewers ever say what the word means. Not that the word is meaningless; rather, it has become a chapter-heading word: a word which stands for a whole range of topics and issues rather than for one definite subject. In particular, the following facts are commonly cited as examples of "intentionality": (1) the fact that words, sentences, and other "representations" have *meaning*; (2) the fact that representations may *refer* to (i.e., be true of) some actually existing thing or each of a number of actually existing things; (3) the fact that representations may be *about* something which does *not* exist; and (4) the fact that a state of mind may have a "state of affairs" as its object, as when someone says, "she believes that *he is trustworthy*," "he hopes that *his boss will get fired*," "she fears that *there won't be food in the house*."

When the computer revolution burst upon the world, it was widely expected (and claimed) that computer models would explain the nature of these various phenomena. In short, people expected that a reductive account of the various topics included under the chapter heading "intentionality" would be given. Now that this has proved not so easy, a number of thinkers are beginning to suggest that it isn't so bad if this can't be done; intentionality is only a feature of "folk psychology" anyway. If a first-class scientific account of intentional facts and phenomena can't be given, that is not because scientific reductionism is not the right line to take in metaphysics, but rather it is because there is, so to speak, nothing there to reduce. I want to argue that both attitudes are mistaken; that intentionality won't be reduced and won't go away.

That claim—the claim that 'intentionality won't be reduced and won't go away"—has sometimes been called "Brentano's thesis," after a philosopher who is (not completely accurately)[1] credited with defending it in the latter part of the nineteenth century. Sometimes the view is stated as a positive claim: the claim that intentionality is a *primitive phenomenon,* in fact *the* phenomenon that relates thought and thing, minds and the external world.

In a sense, this positive view follows immediately from the negative one, but there is a joker in the pack. The joker is the old philosophical problem about the One and the Many. If one assumes that whenever we have diverse phenomena gathered together under a single name, There Must Be Something They All Have in Common, then indeed it will follow that there is a single phenomenon (and, if it is not reducible, it must be "primitive") corresponding to intentionality.

To see the difficulty, consider the property "red." Intuitively, "red" things do all have "something in common." But scientifically, they do not, unless it be a "reflectancy"—a disposition to selectively emit and absorb certain wavelengths of light. Such a disposition—a disposition to affect things in a certain way (things other than human mental states)—would have been called a "tertiary property" in the seventeenth century. (Secondary properties were dispositions to affect our minds, and primary ones were "in the thing itself" just as we conceive them.) If we confine ourselves to nondispositional (or "structural") properties describable in physical science, then there is no scientifically describable property common to all red things, no structural property that constitutes the "redness" of a red star, red light, a red apple, etc., unless we are willing to consider a huge (possibly an infinite) disjunction of structural properties to be a single "physical property." Yet, there is a sense in which they still have "something in common"—something nondispositional and nondisjunctive. In the ordinary-language sense of the term *red,* they are all "red." Of course, whether one admits that this is *really* "something in common" will depend on whether one believes that the commonsense version of the world is just as legitimate as the scientific version. A philosopher who does (like myself) need not give up the claim that red things *do* "have something in common." But he must separate this question from the question "Do they have something in common which is describable in nondispositional terms at the level of exact science?" Things can have something in common in one description of the world and not in another.

I shall try to show that there is no scientifically describable property that all cases of any particular intentional phenomenon have in common. By this thesis I mean to deny that there is some scientifically describable "nature" that all cases of "reference" in general, or of "meaning" in general, or of "intentionality" in general possess; I also mean to deny that there is any scientifically describable property (or "nature") that all cases of any one specific intentional phenomenon, say, "thinking that there are a lot of cats in the neighborhood," have in common. But these phenomena cannot be dismissed as mere folk psychology, unless the very idea that *there are things and we think about them* can be dismissed as folk psychology.

The comparison of intentionality with "red" is misleading, however. A better comparison—one suggested by Wittgenstein[2]—is with the term "game." Even at the ordinary-language level, it is strange to say that all games "have something in common," namely *being games*. For some games involve winning and losing, others ("Ring a Ring o' Roses") do not; some games are played for the amusement of the players, others (gladiatorial games, professional games) are not; some games have more than one player, others do not; and so on. In the same way, when we examine closely all the cases in which we would say that someone has "referred to" something (or even all the cases in which someone has "referred to" one particular thing), we do not find any *one* relation between the word and the thing referred to.

If Wittgenstein says that the word "game" does not stand for a property, there is an obvious criticism that someone can make of his claim, however. "Hume already distinguished between an ordinary (or, as he said, a 'natural') sense of words like 'relation' and 'property' and a logical (or, as he said, a 'philosophical') sense of these words. Why should we regard it as particularly important that there is no property *in the ordinary-language sense of the term* that all games have in common? In the logical sense of the term, there *is* a property all games have in common, namely the *disjunction* of the various criteria that we use to tell things are 'games.'" To this Wittgenstein offers what may look like an unconvincing answer: to represent a "family resemblance" notion like "game" as a disjunction of exact notions is to misrepresent its character. Words like "game" have a vagueness, a flexibility, an "open texture" (as Waismann called it), which no determinate disjunction of completely determinate properties can reproduce.

This looks unconvincing, antiphilosophical, because, after all, isn't the whole *purpose* of rational reconstruction to "tighten up" our vague commonsense notions? We don't want a "rational reconstruction" of a notion to share the vagueness of the preanalytical notion itself. If we remember that it is not words like "game" that Wittgenstein is really interested in, but precisely words like "reference," "language," "meaning," then the situation is very different. Here, I shall try to show, the phenomenon of open texture runs far beyond the mere looseness of conventional application we find in the case of the word "game." (Actually, it runs far beyond mere *looseness* of application in the case of "game" too.) And it is precisely the open texture of reference that defeats the classical philosophical pictures.

If this is right, we have to learn to see that we are in a position which fits neither the philosophical picture of intentionality as a phenomenon to be reduced to physical (or, perhaps, computational) terms, nor the picture of intentionality as a myth, nor even the picture of intentionality as a single "irreducible phenomenon."

Part of my aim is to illustrate (by applying it to a particular problem) a philosophical attitude that gives up many traditional assumptions about Appearance and Reality; that gives up, for example, the assumption that what is real is what is "under" or "behind" or "more fundamental than" our everyday appearances, that gives up the assumption of The One in the Many, and that also gives up the assumption that every phenomenon has an "ultimate nature" that we have to give a (metaphysically reductive) account of. The thrust of my argument is thus negative. I am arguing that a certain way of thinking about meaning and about the nature of the mind is fundamentally misguided. It is always less exciting to hear someone criticize attempted solutions to a problem than to hear him announce that he has found the solution. But I think we can learn something about the nature of meaning and, perhaps, something about the nature of psychology by seeing why certain ideas about meaning and its place in the mind don't work.

Fodor and Chomsky

To explain what is wrong with the way philosophers and cognitive scientists have generally approached questions about meaning, it will be necessary to examine a number of different ways in which the standard approach has manifested itself, a number of different (though intimately related) ways of thinking. The way of thinking I am going to discuss first of all is expounded by Jerry Fodor in *The Language of Thought*.[3] Fodor acknowledges that he owes a great deal of his inspiration to the work of Noam Chomsky. However, Chomsky has never committed himself to the possibility of finding "psychologically real" entities which have enough of the properties we preanalytically assign to "meanings" to warrant an identification. The "representations" and "innate ideas" of which Chomsky writes are deep syntactic structures and syntactic universals. Fodor's program is thus not identical with Chomsky's but rather a daring extension of it.

In any case, there is a widespread expectation that Chomsky's ideas will sooner or later *be* extended to the realm of semantics, an expectation which is responsible for much of the attention that is paid to his ideas by French neostructuralists American cognitive scientists, and others. Chomsky is famous for having proposed a theory according to which grammar is "innate" in the mind. According to Chomsky, there is a Universal Grammar—a structure and a set of categories which are universal, and not just because human environments are in certain respects all alike, but because this Universal Grammar is built into the basic structure of the mind itself.[4] Chomsky further suggests that this innate linguistic structure characterizes not the whole mind, but the way of functioning of a particular "module" in the mind, the so-called "language organ."[5] Chomsky appears to conceive of the language organ as a relatively "dumb" organ, independent of general intelligence ("if there is such a thing," Chomsky would say). This stress on the dumbness of the language organ seems to be a sharp turn from Chomsky's earlier model of the mind as learning its native language—with the aid, of course, of its knowledge of Universal Grammar—by *hypothesis formation*. The more recent writings of Chomsky and Fodor picture the mind as a collection of automatically functioning "modules,"[6] and these writings stress "bottom-up" as opposed to "top-down" processing—that is, automatic processing as opposed to processing which draws on general intelligence and general information.

At any rate, given that the key ideas of Chomsky's theorizing are (1) the idea of Linguistic Universals, (2) the Innateness Hypothesis, and (3) the recent idea of modularity, the form that one can expect a Chomskian theory of the semantic level to take is relatively clear (and Fodor's theory does take the expected form), even if the details can take various shapes. A Chomskian theory of the semantic level will say that there are "semantic representations" in the mind/brain; that these are innate and universal; and that all our concepts are decomposable into such semantic representations. This is the theory I hope to destroy.

I am also skeptical about the idea of Universal Grammar,[7] but I am not going to discuss that in the present work. Chomsky's work, and especially his revival of "mentalism" and his talk of universals in language, has excited worldwide attention, and this is not because people have a tremendous interest in grammar. These ideas have caught the attention of people very far from any concern with technical linguistics: Lacanian psychoanalysts, anthropologists, child psychologists, philosophers of all kinds. Obviously people do anticipate that Chomsky's idea will have implications with respect to issues larger than how we acquire syntax.

I would not try to destroy the theory of innate semantic represen-

tations if I did not think that there is much to be learned from study-
ing the questions it raises and the answers it proposes, and if I did
not think that the brilliant thinkers who propound such theories are
in the grip of an intellectual yearning which is itself worth taking
seriously. The yearning is one which is explained by two facts about
recent thinking about the mind.

One fact is the robustness of the oldest pattern of explanation of
our mental workings there is: explanation in terms of beliefs and desires. No matter how strongly the tides of behaviorism have run, we
have never stopped explaining our behavior and that of others in
terms of beliefs and desires. We say, "I went to school today because
I knew I had to teach a class," or, "I went to the market because I
knew we were out of milk, and I wanted milk to put in my coffee."

Behaviorism in its radical form suggested that we don't need any
of this, because all we are really talking about is conditioned responses, etc. Perhaps one *can* do without belief-desire talk when one
is dealing with rats in very controlled situations, but even so great a
behavior scientist as Skinner ran into trouble when he tried to use
stimulus-response language to describe human verbal behavior.
What Skinner had to do, basically, was to widen the notions of stimulus and response so that (as Chomsky pointed out in a famous review many years ago)[8] they became empty. For example, in the
course of trying to analyze an utterance about World War II, Skinner
referred to the war as the stimulus. Chomsky pointed out that once
the notion of a stimulus becomes so wide that *World War II* is a "stimulus" (and the response takes place twenty years later), stimulus-
response talk has become a mere jargon with no real control. So there
are certainly some good reasons for wanting to defend belief-desire
explanation.

The other fact is the increasing tendency to think of the brain as a
computer and of our psychological states as the software aspect of
the computer. In research based on such an approach (in artificial-
intelligence work, for example) it is often assumed that the computer
has a built-in (and thus "innate") formalized language which it can
use as both a medium of representation and a medium of computa-
tion. (The idea of a *lingua mentis,* a language of the mind, is really an
ancient idea that has made a reappearance, somewhat like the idea
of a Beginning of the Universe.)

If we identify the computer's *lingua mentis* with Chomsky's "semantic representations," we arrive at a familiar picture: the picture of the
mind as a Cryptographer. The mind thinks its thoughts in Mentalese,
codes them in the local natural language, and then transmits them
(say, by speaking them out loud) to the hearer. The hearer has a Cryp-

tographer in his head too, of course, who thereupon proceeds to decode the "message." In this picture natural language, far from being essential to thought, is merely a vehicle for the communication of thought.

The idea of reviving belief-desire psychology and the idea of a computational model of the mind can appeal for many reasons. If Chomsky is right, all mankind has a single nature, just as eighteenth-century thinkers believed. Chomsky has stressed this connection with the Enlightenment, and with the political ideals of Liberty, Equality, and Fraternity.[9] But even apart from this reverberation, it is understandable that many thinkers should feel attracted to a program which brings together belief-desire psychology and computational modeling. The desire to bring these two together gripped me too for a long time. These are the dominant antibehaviorist tendencies, and I believed that they would gain strength by being united.

The desire that grips Fodor, then, as it once gripped me,[10] is the desire to make belief-desire psychology "scientific" by simply identifying it outright with computational psychology. When I proposed this program (under the name "functionalism"), I thought that the way to effect it was simple: we simply postulate that desires and beliefs are "functional states" of the brain (i.e., features defined in terms of computational parameters plus relations to biologically characterized inputs and outputs). For example, one might postulate that *believing there is milk in the supermarket* is displaying one of the formulas[11] in the *lingua mentis* whose translation is "there is milk in the supermarket" in a special "belief register." Displaying another formula in a "desire register" could be *desiring milk for tomorrow's breakfast*. And going from these two computational states to the resultant, i.e., the action of going to the supermarket and buying milk, might be the result of a certain algorithmic procedure on these displayed formulas (as well as on others). In such a picture, ordinary-language mentalistic psychology, folk psychology, is a rough approximation to an ideal computational model of what goes on in the brain. An *ideal* belief-desire psychology would be isomorphic to (a part of) the computational description of what goes on in the brain. Make that assumption and you have *mentalism* in its most recent form.

Mentalism is just the latest form taken by a more general tendency in the history of thought, the tendency to think of concepts as scientifically describable ("psychologically real") entities in the mind or brain. And it is this entire tendency that, I shall argue, is misguided.

Three Reasons Why Mentalism Can't Be Right

1. Meaning Is Holistic

The doctrine called "meaning holism" arose as a reaction to logical positivism; it offered arguments refuting positivist attempts to show that every term we can understand can be defined in terms of a limited group of terms (the "observation terms"). The arguments I refer to were the work of W. V. Quine.[12] These arguments are largely accepted by Fodor,[13] but he does not seem to appreciate their significance for his own enterprise.

Holism is thus, in the first instance, opposed to positivism. The positivist view of language insists that all meaningful descriptive words in our language must have definitions in terms of words in a "basic" vocabulary, a vocabulary which consists of words which stand for notions which are *epistemologically more primitive* than, say, the theoretical terms of science. The favorite candidate of positivists was a vocabulary which consists of sensation terms, or, at any rate, terms for what is supposed to be "observable." If we formulate positivism as a thesis about the truth conditions for sentences rather than as a thesis about the definability of terms, we may say that, as a historical fact, positivists originally insisted that the meaning of a sentence should be given by (or be capable of being given by) a rule which determines in exactly which experiential situations the sentence is assertable.

Now, most of twentieth-century philosophy of science consisted in the gradual overthrow of this view. The logical positivists themselves shifted from advocating the view to criticizing it. Basically, what came to be realized (even by the positivists themselves) was that theories cannot be tested sentence by sentence. If the sentences of which a theory consists had their own independent experiential meanings, or made so many separately testable claims as to what experience will be like, then one could test a scientific theory by testing sentence 1 and testing sentence 2 and testing sentence 3 and so on. But, in fact, the individual postulates of a theory generally have no (or very few) experiential consequences when we consider them in isolation from the other statements of the theory. For example, Newton's Theory of Universal Gravitation (without any added statements specifying boundary conditions) is compatible with any orbits whatsoever. One could even reconcile *square* orbits with the Theory of Universal Gravitation, by saying, "Well, that means there are nongravitational forces acting on the system." It is only in the presence of a large body of statements that one derives all of its so-called "consequences" from a scientific theory. As Quine puts it, sentences meet the test of experience "as a corporate body," and not one by one. (Hence the term "holism.")

The same thing is true of the language of daily life. If someone tells you, for example, that *the thief entered through that window, and there is muddy ground outside the window,* you will "deduce" that *there are footprints in the mud.* But this is not a *logical consequence* of the facts stated, for you obviously have made use of an unstated auxiliary hypothesis to the effect that *if the thief entered through that window, he walked on the ground to get to the window,* and other items of general information as well. If your informant says, "No, he was wearing stilts," then instead of expecting to find shoe prints in the mud you will now expect to find holes of a different shape. What has experiential import is the corporate body of statements, and this import is not the simple sum of the experiential imports of the individual statements.

In ordinary language as opposed to formalized language, this phenomenon is made even more pervasive by what is sometimes called the "nonmonotonicity" of the logic of everyday discourse. In a formalized language, if one says, "All birds fly," and he also says, "Ostriches are birds," one can deduce, "Ostriches fly." But ordinary language isn't like that. If I say, "Hawks fly," I do *not* intend my hearer to deduce that a hawk with a broken wing will fly. What we expect depends on the whole network of beliefs. If language describes experience, it does so as a network, not sentence by sentence.

Meaning holism also runs counter to the great tendency to stress *definition* as the means by which the meaning of words is to be explained or fixed, i.e., counter to that famous stumper "Define your terms!" It has this aspect (which is very much stressed by Quine) because a suggestion that at once emerges from holism is that most terms *cannot* be defined—or, at least, cannot be defined if by a "definition" one means something that is fixed once and for all, something that absolutely captures the meaning of the term.

Why does holism suggest this? Because, when an entire body of beliefs runs up against recalcitrant experiences, "revision can strike anywhere," as Quine has put it. Even if a term is originally introduced into science via an explicitly formulated definition, the status of the resulting truth is not forever a privileged one, as it would have to be if the term were simply a synonym for the *definiens.*

An example from the history of physics may help to clarify this all-important point. In Newtonian physics the term *momentum* was defined as "mass times velocity." (Imagine, if you like, that the term was originally equated with this *definiens* by the decision of a convention of Newtonian physicists.) It quickly became apparent that momentum was a conserved quantity (as Leibniz already thought). With the development of vector analysis, the stereotype of momentum as a quantity which is conserved and which has a scalar value and a direction—the direction of motion of the particle—became universal among physicists. But with the acceptance of Einstein's Special Theory of Relativity a difficulty appeared.

Einstein did not challenge the idea that objects have momentum, or that it is conserved, or that it is in the direction of motion of the particle. But he showed that the principle of Special Relativity would be violated if momentum were *exactly* equal to (rest) mass times velocity.

What to do? Einstein studied the case of "billiard balls" (particles in elastic collision). Since Newtonian physics "works," their momentum must be given by the formula "mass times velocity" *almost* exactly, at least when the velocities are "nonrelativistic" (small compared to c, the velocity of light). Can there be a quantity with the properties that (1) it is conserved in elastic collisions, (2) it is closer and closer to "mass times velocity" as the speed becomes small, and (3) its direction is the direction of motion of the particle? Einstein showed that there *is* such a quantity, and he (and everyone else) concluded that that quantity *is what momentum really is*. The statement that momentum is *exactly* equal to mass times velocity was revised. *But this is the statement that was originally a "definition"!* And it was reasonable to revise this statement; for why should the statement that momentum is conserved not have at least as great a right to be preserved as the statement "momentum is mass times velocity" when a conflict is discovered?

A philosopher of a traditional bent might have answered this last question by saying, "Because 'Momentum is mass times velocity' gives the very *meaning* of the word 'momentum.' You cannot revise an *analytic* truth." But such a philosopher is imposing a set of categories—the ideas of fixed definitions of terms and analytic truths—which have no reality for actual scientific practice. In effect, he treats an accident of history—how the term first came into science—as if it determined the future choices scientists were allowed to make. As Quine puts it, *truth by stipulation is not an enduring trait of sentences.* When the statements in our network of belief have to be modified, we have "trade-offs" to make; and what the best trade-off is in a given context cannot be determined by consulting the traditional "definitions" of terms.

Another traditional move is to say, "Well, the scientists decided to *change the meaning of 'momentum.'*" If this accounts for the change in the truth-value scientists assign to the sentence "Momentum is mass times velocity" after the adoption of Relativity, then it must follow that we are now talking about a different physical magnitude. But no, we are still talking about the same good old momentum—the magnitude that is conserved in elastic collisions. That's the physical magnitude "momentum" always referred to if it referred to anything. And that magnitude, *momentum itself*, turned out not to be exactly equal to mass times velocity.

If this seems strange, it is because we are not used to thinking of meanings as being historic entities in the sense in which persons or nations are historic entities. I, Hilary Putnam, had curly blond hair when I was small. I did not speak English, but only French. I did not think of my name as "Hilary Putnam," but as "Hilaire Poot-nomm." Now I have thinning gray hair, which is not curly at all, I speak English rather than French, and I call myself "Hilary Putnam." Yet I am the same person. There are practices which help us decide when there is enough continuity through change to justify saying that the same person still exists. In the same way, we treat "momentum" as referring to the same quantity that it always referred to, and there are practices which help us decide that there is enough continuity through change to justify doing this. Meanings have an identity through time but no essence.

2. Meaning Is in Part a Normative Notion

I have argued elsewhere[14] that the notions of being a justified or warranted or reasonable belief are not reducible to physicalistic notions. Some of the arguments will appear in later chapters of this book. But even if one could give a reductive analysis of the notion of being a justified belief, say, by identifying "being justified" with "being the outcome of such and such methods," or such and such an algorithm, or such and such a computer program, that algorithm would have to be as complex as a description of the "general intelligence" of an idealized inductive judge. We have seen, from our brief discussion of meaning holism, that testing a scientific theory is not something that can be done just by looking up the operational definitions of all the terms and testing the sentences that comprise the theory one by one. Rather, it involves very intangible things, such as estimating simplicity (which itself is not a single factor, but different things in different situations), and weighing simplicity against our desire for successful prediction and also against our desire to preserve a certain amount of past doctrine. It involves having a nose for the "right" trade-off between such values. The ability to make these estimates and trade-offs is what Fodor calls "general intelligence," and he does not expect general intelligence to be explained in terms of "modules" in the foreseeable future, if ever. Describing the nature of general intelligence is a hopeless problem, according to Fodor, and the whole point of Fodor's "modularity hypothesis" is to *separate* the problem of understanding the "language organ" from the problem of understanding general intelligence.

Now, I want to say that the notions collected under the chapter heading "meaning" (or "intentionality"), for example the crucial notions of "same meaning" and "same reference," are as complex as the notions collected under the chapter heading "general intelligence." This is not to claim that it *always* requires a great deal of intelligence to tell that two terms have the same meaning or the same reference. But there are many cases in which it doesn't require a great deal of intelligence to solve a problem in inductive or deductive reasoning. To determine the intrinsic complexity of a task is to ask, *How hard can it be in the hardest case?*

Well, a theory of synonymy would be a theory that decided questions of interpretation. Consider, however, just how subtle questions of interpretation can be, even when we deal with texts that aren't particularly "literary." The fact that scientists who used the word "momentum" were using it as a name for a conserved quantity rather than as a synonym for "mass times velocity" (even if they called that "the definition of momentum") has already been mentioned. Another example is our knowledge of the fact that when Bohr used the word "electron" (*Elektron*) in 1934, he was talking about the same particles he called "electrons" in 1900. We do not determine this by comparing the *theories and descriptions* of the electron that Bohr gave at these two different times and seeing that they were much the *same,* because they *weren't.* The 1900 theory said that electrons go around the nucleus just as planets go around the sun, i.e., electrons have trajectories, whereas the 1934 theory (which is, in essence, the present quantum theory) says that an electron never has a trajectory—in fact, it never has a position and a momentum at the same time. Yet a physicist might well describe the development of the later theory from the earlier in this way: in the nineteenth century we discovered that electrons have a certain mass-charge ratio by deflecting electron beams in a magnetic field; later we discovered by another experiment what the electron charge is (and hence what the value of the electron mass must be); we discovered that electric current is a stream of electrons; we discovered that every hydrogen atom consists of one electron and one proton; we thought for a time that electrons had trajectories, but then we discovered the Principle of Complementarity; and so on. In short, he would tell the story as a story of successive changes of belief about the same objects, not as a story of successive "changes of meaning." And the same kind of "general intelligence" is involved in his decision to treat all of these occurrences of "electron" as synonymous as is involved in his decision to treat later research programs in the story as extensions of the earlier ones; a kind of decision that plays a central role in theory evaluation. In fact, treating "electron" as preserving at least its *reference* intact through all of this theory change and treating Bohr's 1934 theory as a genuine *successor* to his 1900 theory are virtually the same decision: the decision described once as a decision about the meaning or reference of a term and once as a decision about the familial relations of research programs.

This decision illustrates what has been called "charity" or "benefit of the doubt" in interpretation.[15] When we interpret Bohr in 1900 as referring to what *we* call "electrons," we are thereby making at least some of his 1900 beliefs come out "true" by our lights, whereas interpreting him as referring to nonexistent objects would be to dismiss all of his 1900 beliefs as totally wrong. And, of course, Bohr in 1934 extended the *same* "charitable" attitude toward his 1900 self that we do (which is why he continued to use the word "electron" in all those papers.)

All interpretation depends on charity, because we always have to discount at least *some* differences in belief when we interpret. For example, suppose we are reading a novel written two hundred years ago in English, and we encounter the noun "plant." In a normal context, we do not hesitate to identify this "plant" with our present English "plant"; yet, in so doing, we are ignoring a host of differences in belief. For example, *we* believe that plants contain chlorophyll, we know about photosynthesis and the carbon dioxide–oxygen cycle, and so on. These things are central to our present notion of what a plant is. All of these things were unknown two hundred years ago. Yet (unless we are philosophers or philosophically minded historians of science) we do not say that people two hundred years ago "lived in a different world," or that their notions are "incommensurable" with the notions we now have,[16] which taken literally (of course, it never is!) would imply that we could not interpret an ordinary letter that anyone wrote two hundred years ago. In short, we treat the concept *plant* as having an identity through time but no essence, and we treat the concept *electron* as having an identity through time but no essence.

And yet, we do not always interpret words in such a way as to *maximize* the number of true beliefs that the speaker would have had (by our lights) if the interpretation were correct, contrary to a crude version of the idea of "charity in interpretation." Here is a counterexample to this crude version: the great metallurgist Cyril Stanley

Smith once proposed to me (as a joke, but one with a serious point) that there really is such a thing as *phlogiston* (the substance that was supposed, before the role of oxygen was discovered, to account for combustion by *leaving* the burning substance and gradually saturating—or "phlogisticating"—the air). Phlogiston, Smith suggested, is *valence electrons*. What makes this a joke is that, as Smith perfectly well knows, we do not speak as he "proposed" we should; we are not prepared to say, "Phlogiston theorists were talking about valence electrons, but they had some of the properties wrong." That would be excessive "charity." The knowledge that one thing is reasonable charity while another thing would be excessive exhibits our full powers of understanding, whether the context be interpretation or "real life." There is no hope of a theory of sameness of meaning or reference which applies to such difficult cases and which is independent of our account of "general intelligence."

What hangs on these difficult decisions is extremely important to us. It is important to us if we are reading a novel, because the decision to treat the words of the novel as *alien* ("incommensurable") would, were we to make it, utterly change our relation to the literary work. And it is important to us if we are trying to understand the history of science, because the interpretation we give to the scientist's words will play a large role in establishing for us the scientist's achievement or lack of achievement, his rationality or lack of rationality. In this way, deciding to interpret someone one way rather than another is intimately tied to normative judgments.

If we reflect on the role played by the notion of sameness of meaning in logic,[17] it will perhaps not seem so surprising that this notion turns out to have a normative dimension. In logic, *equivocating*, i.e., using a term in one sense at one point in an argument and in a different sense at a different point in the same argument, is a fallacy whether the argument be inductive or deductive. But the notion of "sense" or "meaning" (Fodor's "content") could not play this role in criticism if we did not interpret one another in such a way that "meanings" are preserved under the usual procedures of belief fixation and justification. If we adopted the meaning proposals of operationists or positivists according to which modifying a scientific theory virtually always produces a "change in the meaning" of the theoretical terms, then we would have to say that every scientist who modifies an existing theory in order to solve a problem that someone poses is guilty of *equivocation*. Without a doubt, we would quickly introduce—or rather, *re*introduce—the traditional notion of "change of meaning" so that we could distinguish between cases in which a scientist has committed a real "fallacy of equivocation" in answering a question and cases in which it is only in a Pickwickian sense that a "change of meaning" has taken place between the original posing of the question and the giving of the answer.

In sum, it is essential to the role played by such notions as "equivocation"—it is presupposed by the *epistemic* role that these notions have—that we interpret one another in such a way that the "meaning" of a word is the same (in the sense of being *treated* as the same) under normal procedures of belief fixation and justification. (It is because interpretative practice owes allegiance to this constraint that sameness and difference of meaning cannot coincide with the presence and absence of any local computational relation among our "mental representations." As the previous examples illustrate, a computational relation which coincided with synonymy could not be "modular" in Fodor's sense, that is, could not be psychologically more elementary than "general intelligence.")

3. Our Concepts Depend on Our Physical and Social Environment in a Way That Evolution (Which Was Completed, for Our Brains, about 30,000 Years Ago) Couldn't Foresee
To have given us an innate stock of notions which includes *carburetor, bureaucrat, quantum potential,* etc., as required by Fodor's version of the Innateness Hypothesis, evolution would have had to be able to anticipate all the contingencies of future physical and cultural environments. Obviously it didn't and couldn't do this.

Connections between 1, 2, and 3

Mentalists who follow Fodor's lead are committed to the idea that there is an innate stock of semantic representations in terms of which all of our concepts can be explicitly defined. Point 3 raises an obvious difficulty: *How could such concepts as "carburetor" possibly be innate?* Primitive peoples who have had no acquaintance with internal combustion engines show no difficulty in acquiring such concepts. On Fodor's account this means that their "language of thought" contained the concept "carburetor" *prior to their acquiring a word for that concept,* even though nothing in their evolutionary history could account for how the concept "got there."

A mentalist (like my former self) who is not of the MIT variety may, of course, believe in "computational psychology" *without* accepting the Innateness Hypothesis. But he still faces serious difficulties. It is part of the very notion of a computational psychology that all representations must be described syntactically or procedurally, or by a combination of syntactic and procedural features. At the same time, a large part of the argument for meaning holism is that changes in a community's "procedures" for using a lexical item do *not* usually count as a change in the meaning of the item.

If our thinking is ultimately conducted in an internal language of thought (only not an innate one), the same thing will be true of the items in *lingua mentis* corresponding to the words in public language we have been using as examples ("momentum," "electron," "plant," etc.). Moreover, if the *lingua mentis* is not innate, then the mental representation corresponding to a given item in public language may vary, at least in syntactic respects. Germans not only may use a different word for *plant* than English speakers; they also may use a different "mental word," if the mental vocabulary is not innate. Words in the *lingua mentis* of different speakers which have different syntactic "shapes" (different "spellings," so to speak), and different "procedures" associated with them, may actually have the same meaning and denotation. (If we just recommit the noise "meaning" so that any difference in the procedures one speaker associates with the word "cat" and the procedures a second speaker associates with "cat" count as a difference in the "meaning" of the word, then we will not have a theory of meaning, but just a complete change of topic.) In addition, if semantic representations in the brain are developed from experience, just as words in a public language are, rather than being composed out of an innate set of semantical primitives, there is no reason to think that a given representation (described syntactically) will not come to be given *different* meanings by different groups of human beings. ("Different meanings" by the criteria used by a good interpreter, this means.) A word in *lingua mentis* might, so to speak, have one meaning for French speakers and a different meaning for English speakers, just as a written word sometimes has one meaning in French and a different meaning in English. "Translating" our public language into *lingua mentis* will not solve the problem of conceptual content, but only move it from one language to another. I shall not dwell further on these problems now, but they will reappear in future chapters.

There are connections between 1, 2, and 3 which it is important to see. The argument against positivism and against the possibility of defining all of our concepts from some basic stock of "epistemologically primitive notions" was summarized under point 1. The heart of the argument was that to adopt a notion of "meaning" according to which ordinary scientific discoveries (discovering that water is H_2O, that momentum is not exactly the product of mass and velocity, that electrons obey the Principle of Complementarity, or that plants contain chlorophyll and perform photosynthesis) *change the meaning* of the relevant terms would violate the principle mentioned under point 2, the principle that meanings are to be treated as the same under ordinary processes of belief fixation and justification. To say that we changed the meaning of the word "water" when we decided that water is H_2O would not only go against our intuitions of synonymy; it would violate this interpretative principle, which is central to the epistemic function of the notion "change of meaning."

In addition, there is a connection between points 1 and 3: if the early views of the logical positivists had been right (i.e., if point 1 had been false), then evolution would not have had to give us such unlikely "innate" concepts as *carburetor* or *positive charge*, even if the Innateness Hypothesis were true; it would only have had to give us some stock of basic notions (the observation terms) from which they could be defined. But (as Fodor recognizes) our terms cannot be defined from a set of terms much smaller and biologically more primitive than the whole lexicon. In short, the truth of meaning holism blocks the only way of meeting objection 3 that makes biological sense.[18] (In *The Language of Thought* Fodor does not try to answer objection 3; instead he simply marvels at the fact that all these unlikely concepts must be innate—since that is required by the facts, on his theory.) In sum, sophisticated mentalism of the MIT variety is not blocked by any one of these points separately, but by 1, 2, and 3 acting together.

In fact, my real reason for beginning our discussion of computationalist (and physicalist) theories of meaning with an examination of the theory of *The Language of Thought* was precisely to illustrate the way in which theories are likely to run afoul of the principle of meaning holism and to run afoul of various "principles of charity" (in particular, the principle that meanings are preserved under normal belief fixation). I don't think that Fodorians and Chomskians are a majority among cognitive scientists or philosophers who favor computationalist and physicalist theories of meaning; but the arguments I have deployed against Fodor, especially those based on meaning holism and on the interpretative maxim that meanings are not altered by ordinary procedures of belief change and justification, will reappear when we consider theories which are not committed to Chomsky's Innateness Hypothesis.

In subsequent chapters there will be a number of other issues we shall have to discuss as well. To introduce some of these issues, we must now consider an important aspect of meaning that I have so far deliberately neglected in this discussion. This is the way in which meaning is "interactive," that is, the way in which it depends not just on what is in our heads but also on what is in our environment and on how we interact with that environment. This will be the subject of the next chapter.

Chapter 2
Meaning, Other People, and the World

As he was the first to theorize in a systematic way about so many other things, so Aristotle was the first thinker to theorize in a systematic way about meaning and reference. In *De interpretatione* he laid out a scheme which has proved remarkably robust. According to this scheme, when we understand a word or any other "sign," we associate that word with a "concept." This concept determines what the word refers to. Two millennia later, one can find the same theory in John Stuart Mill's *Logic*, and in the present century one finds variants of this picture in the writings of Bertrand Russell, Gottlob Frege, Rudolf Carnap, and many other important philosophers. Something like this picture also appears to be built into the English language. Etymologically, *meaning* is related to *mind*. To *mean* something was probably, in the oldest usage, just to *have it in mind*. Be this as it may, the picture is that there is something in the mind[1] that picks out the objects in the environment that we talk about. When such a something (call it a "concept") is associated with a sign, it becomes the meaning of the sign.

This picture, whether we trace it back to Aristotle or to the metaphysics built into our language, is worth looking at closely. Let us write down the assumptions that constitute the picture for the purpose of inspection. (In writing them down, instead of the word "concept" I shall use the currently popular term "mental representation," because the idea that concepts are just that—*representations in the mind*—is itself an essential part of the picture.)

1. Every word he uses is associated in the mind of the speaker with a certain mental representation.

2. Two words are synonymous (have the same meaning) just in case they are associated with the *same* mental representation by the speakers who use those words.

3. The mental representation determines what the word refers to, if anything.

These assumptions are likely to seem self-evident, but I believe that they are in fact *false*, and it is necessary to appreciate the extent to which they are false before we can make progress in any discussion having to do with meaning or mental representation.[2]

To say that they are false is to say that there cannot be such things as "mental representations" which simultaneously satisfy all three of these conditions. I do not deny that there are, in some sense, mental representations. We often think with the aid of words and pictures and other signs, and it may be that unconscious thought is even richer in the use of representations than we know. Certainly computational models of the mind/brain rely heavily on the idea of processing representations. But remember that the Aristotelian theory of meaning with which we have been stuck these two thousand–plus years doesn't *just* say that we think in terms of mental representations. It is essential to the theory that sameness and difference of these representations is what *sameness of meaning* is about; that when we say that two words do or do not have the same meaning, what we are saying is that they are or are not associated with the same mental representation. It is also part of the Aristotelian picture with which we have been stuck these two thousand–plus years that sameness and difference of the associated mental representations is what determines whether two words do or do not *refer* to the same things. Both of these latter assumptions, I shall argue, are false.

A way of seeing what is at issue, perhaps, is this: the Aristotelian model is what I spoke of (in the last chapter) as a Cryptographer model of the mind. Everyone recognizes that sameness and difference of meaning are not the same things as sameness and difference of *word* (or sign). The French word *chat* is not the same word as the English word *cat*, but the two words have the same meaning. Again, sameness and difference of reference are not the same things as sameness and difference of word (or sign). Phonetically, at least, "he" is the same sign in Hebrew and in English; but in Hebrew "he" means *she!* Again, "bonnet" is phonetically (and in spelling) the same word in American English and in British English, but in British English "bonnet" can denote the hood of a car, whereas it cannot in American English. Moreover, Hebrew "he" and English "he" are both personal pronouns, and (of course) American "bonnet" and English "bonnet" are both concrete nouns. In each case the two words are indistinguishable at the level of *syntax*. So A and B can be syntactically and phonetically the same word in two different languages (or in two different dialects or idiolects of the *same* language) and yet have different reference. Conversely, there are, of course, many examples of words with different phonetic shape but exactly the same reference.

These things are so obvious that no thinker has ever supposed that sameness and difference of meaning are the same thing as sameness and difference of the syntactic properties (including spelling and phonetic shape) of the sign. But the Cryptographer model—the model of sign understanding as "decoding" into an innate *lingua mentis*—postulates that at a deeper level there is an identity between sign and meaning (this is the fundamental idea of the model, in fact). The idea is that in the *lingua mentis* each sign has one and only one meaning. Two words in human spoken or written languages which have the same meaning are simply two different "codes" for the same item (the same "concept") in the *lingua mentis.*

Even in the *lingua mentis,* on the other hand, it is supposed to be possible for two different representations to have the same *reference* (denotation). For example, "rational animal" and "featherless biped" are two different "concepts" which have the same reference (a popular example of Greek philosophers). But each sign in *lingua mentis* picks out a set of things and it picks it out unambiguously in each possible world. In some versions of the theory, what makes the concepts *rational animal* and *featherless biped* different concepts, even though the same things fall under both of them, is simply that there is *some* possible world in which there are rational animals which are not featherless bipeds and/or featherless bipeds which are not rational animals. Thus the *lingua mentis* is pictured as a kind of Ideal Language in which different signs always differ in meaning and in which different signs also differ in reference, not necessarily in the actual world, but at least in some possible world. If we succeed in decoding a message sent in our local natural language back into the *lingua mentis,* then by inspecting the resulting "translation" (in "clear," as cryptographers say) we shall see at once which words in the message have the same meaning and which have different meanings, which words have the same reference in all possible worlds and which words differ in reference in at least some possible worlds.

By this point we should be quite suspicious. What makes it plausible that the mind (or brain) thinks (or "computes") using representations is that all the thinking we know about uses representations. But none of the methods of representation that we know about— speech, writing, painting, carving in stone, etc.—has the magical property that there *cannot be* different representations with the same meaning. None of the methods of representation that we know about has the property that the representations *intrinsically* refer to whatever it is that they are used to refer to. All of the representations we know about have an association with their referent which is contingent, and capable of changing as the culture changes or as the world changes. This by itself should be enough to make one highly suspicious of theories that postulate a realm of "representations" with *such* unlikely properties. (As we shall see, the mental representations postulated by Fodor and Chomsky do not have property 3. Thus they avoid some of the problems avoided by the traditional view. But not all—the same representation always has the same "content" on their view, even when the speakers have grown up in radically different environments. What problems this poses for their view is a topic to which I shall return.)

I already suggested in the last chapter that if there were a *lingua mentis* and we *could* translate our local natural language into it, we would not have solved any of the problems connected with meaning or reference; precisely the same problems would simply rearise for the *lingua mentis* itself. In particular, I want to argue that to the extent that we *do* think using mental representations, those representations cannot satisfy assumptions 1, 2, and 3 above.

The Division of Linguistic Labor

The word "robin" does not refer to the same species of bird in England and in the United States. (Neither does the word "sparrow.") Suppose that you are an American who is unaware of this fact, and you simply know that "Robins have a red breast." Suppose Jones is an Englishman who is unaware of this fact, but who also knows that robins have red breasts. Then Jones and you may very well be in the same mental state in all semantically relevant respects with respect to the word "robin." Every neurological parameter that could have anything to do with fixing the way you understand the word "robin" may have the same value in your brain and in Jones's brain. Yet the word simply does not refer to the same species on your lips and on Jones's lips. The mental representation associated with the word "robin" may be the same in Jones's brain (or in Jones's mental imagery, etc., if you do not wish to assume that this is reducible to something in the brain), yet the reference is not the same. If there is a word, say "ZYX," associated with "robin" in your *lingua mentis*, then "ZYX" has a different extension in your *lingua mentis* (call it "American deep English") and in Jones's *lingua mentis* (call it "British deep English"). Moreover, the reason is not hard to explain. *Reference is a social phenomenon.* Individual speakers do not have to know how to distinguish the species Robin from other species reliably, or how to distinguish elms from beeches, or how to distinguish aluminum from molybdenum, etc. They can always rely on experts to do this for them. Even in the case of so important a metal as gold, the average person is highly unreliable (in distinguishing gold from brass, etc.) and knows that he is unreliable. That is why he goes to a jeweler (or even to a chemist or a physicist) if he has to "make sure" that some item really is gold.

Let us stick to the word "gold" for a moment. Gareth Evans[3] suggested that the average man doesn't really know the meaning of such words as "gold," that he only knows part of the meaning of such words. But what then is the whole meaning of the word "gold"? Is the whole meaning of the word "gold," "Element with atomic number 79"? This would be a fantastic theory. Has any linguist or philosopher ever suggested that it is *analytic* that gold has atomic number 79? In point of fact, if we should find out that some incredible scientific error has been made, and that the atomic weight of the metal jewelers and ordinary people call "gold" is not 79, we would not say that that metal wasn't really gold, but we would say that gold didn't have the atomic number 79. The chemist who knows that the atomic number of gold is 79 doesn't have a better knowledge of the *meaning* of the word "gold," he simply knows more *about* gold.

What of jewelers, metallurgists, and so on? They know a variety of tests by which they can tell whether or not something is gold. In Locke's time, a favorite test involved being "soluble in aqua regia" (a weak solution of nitric acid, I believe). Is it possible that it is the jewelers who know the whole meaning of the word "gold," and that laymen (and even the scientists, who know the atomic number but don't know the tests used by jewelers) know only *part* of the meaning of the word "gold"? But what if the tests used by jewelers are not the same in different parts of the United States, or if they are not the same in the United States and in England, or not the same in different decades? If jewelers on the West Coast are acquainted with one test for being gold and jewelers on the East Coast are acquainted with a different test, we wouldn't conclude that the word "gold" had one *meaning* on the East Coast (known, in full, only to jewelers on the East Coast) and a different *meaning* on the West Coast (known, in full, only to jewelers on the West Coast).

In any case, the move of saying that the *whole* meaning of the word "gold" is known only to some group of experts, however we decide which group that is, and of saying that the rest of us know only part of the meaning of the word "gold," is not available to mentalists (although Gareth Evans would have disagreed).[4] For the whole aim of mentalism is to identify the meaning of a word with something that is in the brain/mind of *every* speaker who knows how to use the word. It is a constraint on mentalistic theories of meaning that meanings must be *public*. A theory of meaning which makes meaning the, so to speak, property of a group of experts would not explain what thinkers like Fodor and Chomsky want to explain.

What is going on here? If different experts are acquainted with different criteria for being gold, and the person on the street is not acquainted with any very good criterion at all, but has to rely on the experts, then how can we even speak of the word "gold" as having a meaning?

According to the view that I have put forward,[5] the answer to this question has two parts. First, what is in people's brains or minds, their mental representations or mental descriptions or mental pictures, does not in general determine the reference of a word that they know how to use. In the case of most of us, our mental representation doesn't do much beyond telling us that gold is a yellow precious metal to help determine the reference of the word "gold." It certainly doesn't pick out the reference of the word "gold" exactly. In the case of "sparrow" or "robin" the mental representation does even less, and in the case of "elm" and "beech" the mental representation is hopeless (at least if it's *my* mental representation). But what this shows isn't that these words fail to refer, but that the mental representation isn't what picks out their reference, or at least that the mental representation of the typical speaker isn't what picks out their reference. As long as we stick with Aristotle's assumption that the word "hooks on to the world" because it is associated with a mental representation which hooks on to the world, we will be blind to facts which are, so to speak, under our noses. We will keep thinking that the mental representation *must* pick out the referents of the word, because if it doesn't then what *could*? If we have equated the mental representation with the "meaning" of the word from Square One, then we shall simply take it for granted that the *meaning* of a sign must simultaneously (1) be something mental and (2) "hook on to the world." (As Wittgenstein often pointed out, a philosophical problem is typically generated in this way: certain assumptions are made which are taken for granted by *all* sides in the subsequent discussion.)

Suppose we abandon these assumptions. Then we are free to grant that reference exists and is important and interesting, and that mental representations exist and are, perhaps, important and interesting, but we don't have to *identify* problems of reference and problems of mental representation any more. (As I mentioned above, this is a point of which Fodor and Chomsky are perfectly aware.)

Let us look and see what happens if we separate the problems. To begin with, let us look at the problem of reference. We shall see later that it is difficult—I suggest, in fact, impossible—to give a *reductive* theory of reference. But if what we ask is not a reduction of the notion of reference to other notions regarded as metaphysically more basic, or a theory of "how language hooks on to the world," but simply a working characterization of how it is that words like "robin" and "gold" and "elm" manage to refer, then it is not difficult to give one. The fact is that some people know a good deal about certain kinds of things. These "experts" as I have been calling them may pick out these classes by different criteria. That doesn't matter as long as the criteria in fact pick out the same class. If experts in one country determine whether something is gold by seeing whether it is soluble in aqua regia and experts in another country determine whether it is gold by seeing whether it passes some other test, provided the two tests agree (or agree apart from borderline cases), then communication can proceed quite well. There is no reason to think of any one test as "the meaning" of the word. Indeed, the very same community may change from one test to another without anyone being aware of this (each expert may be unaware that almost all of the other experts have changed over to the new test).

But, it will be objected, this only accounts for how experts can use the word. However, there is no problem about how nonexperts can use the word: in doubtful cases, they can always consult the local experts! There is a *linguistic division of labor.* Language is a form of cooperative activity, not an essentially individualistic activity. Part of what is wrong with the Aristotelian picture is that it suggests that everything that is necessary for the use of language is stored in each individual mind; but no actual language works that way.

In sum, reference is *socially* fixed and not determined by conditions or objects in individual brains/minds. Looking inside the brain for the reference of our words is, at least in cases of the kind we have been discussing, just looking in the wrong place.

(If this is accepted, then a new puzzle may arise: why have a notion of meaning at all? If we can account for how our words refer to the things they do without appealing to the idea that they are associated with fixed "meanings" which determine their reference, then why should we have such a notion as meaning at all? But this is not really such a puzzle: the best way to get along with people who speak a different language—or, on occasion, even to get along with people who speak one's "own" language in a different way—is to find an "equivalence" between the languages such that one can expect that— after due allowance for differences in beliefs and desires—uttering an utterance in the other language in a given context normally evokes responses similar to the responses one would expect if one had been in one's own speech community and had uttered the "equivalent" utterance in one's own language. As a "definition" of sameness of meaning this would not satisfy a skeptical philosopher like Quine: it would not satisfy him because, for one thing, the identification of contexts as "the same" presupposes the very "translation scheme" which is being tested for adequacy, and because the identification of beliefs and desires likewise presupposes translation. But in the real world, our problem is not the theoretical problem of "underdetermination"—the problem of the existence of alternative schemes which satisfy the criterion of adequacy equally well—but the difficulty of finding even *one* which does the job. That we do succeed in finding such schemes in the case of all human languages is the basic anthropological fact upon which the whole notion of "sameness of meaning" rests.)

Elms, Beeches, and Searle

John Searle[6] has vigorously attacked the above argument. (In discussion,[7] however, he has admitted that mental representations do not satisfy assumption 2, above. Thus his attack is not incompatible with my position, although his writings suggest the contrary.) What he defends are assumptions 1 and 3. He contends that we have mental representations which determine the referents of common nouns, pronouns, and so on. Why does the argument from the division of linguistic labor not refute this? According to Searle, the way in which I am able to have a representation of elms which does in fact single out elms from all other species, even though I cannot identify elms, is this: my own personal "concept" of an elm is simply *tree which belongs to a species which experts on whom I rely (at this time) call by the name "elm."*

Searle does not, of course, claim that people consciously (or even unconsciously) think, "When I say 'elm' I intend to refer to the trees which experts on whom I rely at this time call by the name 'elm.'" What he believes is that this is their "intention" whether they formulate it to themselves in words (or unconscious representations) or not. That there are "intended" conditions of reference is a fundamental assumption of his theory. Moreover, this claim is accompanied in Searle's writing[8] by a strange metaphysical story about how language hooks on to the world: the capacity of a concept in my mind to refer to something outside my mind is, Searle says, *explained by the brain's chemistry.* For the time being I want to avoid discussing metaphysical questions about how a language-world connection is possible at all;[9]

but some of the considerations I shall raise in the next chapter against the possibility of reducing reference to computational cum physical relations also apply against the possibility of a direct reduction of reference to physics and chemistry of the kind Searle seems to envisage.

As I mentioned, Searle has conceded that concepts, in his sense, cannot be identified with *meanings*. But it is worth seeing why.

No philosopher, certainly not Searle, has ever maintained that it is *analytic* that elms are called by the name "elm" (by English speakers, or by me, or by experts on whom I rely, etc.). For example, suppose the meaning of the word "elm" (in English) were *species of tree which is called by the name "elm" by English speakers* (or by English experts). By parity of reasoning, the meaning of the German word "Ulme" (the German word for elms) must be *species of tree which is called by the name "Ulme" by German speakers* (or by German experts), and the meaning of the French word "orme" (the French word for elms) must be *species of tree which is called by the name "orme" by French speakers* (or by French experts). On this theory, it would be a *mistake* to translate the English word "elm" by the German word "Ulme" or by the French word "orme." Indeed, the three words differ in *meaning*, on this theory, just as much as "elm," "beech," and "maple" do! Moreover, since German has no word for *species of tree which is called by the name "elm" by English speakers*, there would be no way (except by using some such cumbersome locution as *Art von Baum die englisch sprechende Leute «elm» nennen*) to translate the English word "elm." A myriad common English nouns would be translatable into German only with immense difficulty. (And if one did translate, say, an English novel into German using these cumbersome locutions, a German speaker would not be able to understand the result!)

The fact is that *tree that English speakers call "elm,"* or rather *Art von Baum die englisch sprechende Leute «elm» nennen*, is not a *translation* of the word "elm" at all. That elms are called "elms" is not part of the concept of an elm, it is simply something very important to me as an English speaker. Few things could be more important, in fact, to an English speaker who wants to talk about the species than to know its name; but the importance of the fact doesn't make it part of the meaning of the name "elm" that these trees have that name in English. An important part of the purpose of the notion of *meaning* is precisely to *abstract away from* the phonetic shape of the name. To say that the phonetic shape of the name ("elm," or "Ulme," or "orme") is essential to the meaning is to confuse precisely what we want to abstract away from in meaning talk.

Some may retort that meaning talk is, after all, just a piece of folk psychology and we should drop the whole notion, at least in science. Quine has argued such a position for many years. Be that as it may, if we want to give a correct account of the notion of meaning—whether in the end we want to retain it or not—then we have to say that in meaning talk we equate "elm," and "Ulme," and "orme."

In fact, Searle's view has even more radical consequences if we equate his mental representations (or "intensions," as he calls them, using a traditional term for *meanings*) with meanings. As Searle is aware, it is perfectly possible that different English speakers use the word "elm" to refer to different species of tree, without my necessarily being aware of this. (Remember that something similar actually happens in connection with the words "robin" and "sparrow.") For this reason, Searle would say that what I mean (the "intension" with an *s*) when I use the word "elm" is not *tree which belongs to a species which is called "elm" in English* (i.e., by experts about common deciduous trees who speak English), but rather *tree which belongs to a species which is now called "elm" in English by the experts on whom I rely right now.* The reference to *me* is necessary because, as I said, the elms that I am talking about may not be the same as the elms someone else (say, Jones in Nova Scotia) is talking about. My *concept* of an elm (on Searle's theory) is the same as Jones's (just as "I" is the same concept, on Searle's view, whether I think it or Jones thinks it), but the *reference* may be different. Moreover, it may be that at some future time in my life I and the experts on whom I rely will use the word "elm" to refer to a species of tree different from the species we now call by that name. My intention in talking about elms right now (and also the intension of the word, according to Searle) is not to refer to the trees which are called "elms" by any experts at any place in the universe, or by any experts on whom I shall ever rely in my life, but to refer to the trees which are called "elms" now by the experts on whom I am prepared to rely right now. Thus, the intension of the word "elm" must contain both an indexical referring to myself and an indexical referring to the present time.[10]

The point is important enough to deserve restatement: I *can* incorporate my knowledge of the division of linguistic labor into my description of what I am referring to by using a phrase like *species of tree which is called "elm" by such and such experts.* Indeed, one sometimes *has* to fall back on such a description when a word does not have a synonym in the language one is speaking. For example, if there is a kind of bird which is called a *chooc* in a language spoken in the Amazon jungle, say Natool, and I have no name for that species in English, then I may have to explain what is meant by *chooc* (i.e., what sort of thing the word is used to refer to) by saying, "Well, they use the word to refer to a species of bird that they call a *chooc*." But such descriptions as *species of bird that the Natool call "chooc"* do not give us synonyms for the words whose use is so explained; rather, they are a way of bypassing the need for a synonym. Once again what we see is the impossibility of identifying meanings with the descriptions that speakers "have in their heads," i.e., of identifying the notions of *meaning* and *mental representation*.

Against this argument, it is sometimes contended that our mental representations of an elm and a beech *must* be different since we *know* that elms and beeches are different species. This counterargument is, however, fallacious. I do know that elms and beeches belong to different species; thus it is included in my "mental representation" of an elm that it is not a beech (whatever sort of tree a beech may be) and it is included in my "mental representation" of a beech that it is not an elm (whatever sort of tree an elm may be). But what this *amounts to* is that my mental representation of an elm includes the fact that there *are* characteristics which distinguish it from a beech, and my mental representation of a beech includes the fact that there *are* characteristics which distinguish it from an elm. The situation is totally symmetrical. It remains the case that the only difference between my "mental representation" of an elm and my "mental representation" of a beech is my knowledge that the former species is *called* "elm" and the latter species is *called* "beech." Apart from the differences in the phonetic shapes of the names (which, as we have seen, cannot be a part of the *meaning* of the names), there is no difference between my "mental representation" of a beech and my "mental representation" of an elm. Knowing that there are two different "species" is knowing that there *exist* distinguishing characteristics; one can know this without its being the case that those distinguishing characteristics are themselves included in the "mental representations." If the distinguishing characteristics were themselves included in the mental representations, then indeed the representations of an elm and a beech would be different, even apart from my knowledge of the phonetic shapes of the names; but the mere knowledge of the *existence* of (unspecified) distinguishing characteristics does not make the representations different, except in the trivial way mentioned.

A different move, one that I have heard Fodor make many times, is to "bite the bullet" and say that *in one sense of "meaning"*—he calls this sense of "meaning" *narrow content*—the meaning of "elm" and "beech" is exactly the same, and that this sense of "meaning" is the one that is of interest to psychology. But meaning, we should recall, if it is anything, is what we try to preserve in translation. The one thing we *don't* do in translation is translate "elm" as "beech" (or as

"Buche," if we are translating into German). It may be that *something* of psychological interest which is associated with the word "elm" is the same, for example the "stereotype." My stereotype of an elm is that of a common deciduous tree, and this is also my stereotype of a beech. But to call stereotypes "contents" (or "narrow contents") is not to offer a theory of meaning, but rather to change the subject. This is a point to which I shall return. (Note also that stereotypes are not just "images"; they are, at least in part, beliefs stated in *words*. Thus even if we did decide that stereotypes are a "component" of meaning, the identification of this component is parasitic on the *ordinary* notion of "meaning.")

The elm-beech case (and also the case of gold) enabled us to see two things: first, that what is preserved in translation isn't *just* "mental representations," and second, that "mental representations" don't suffice to fix reference.[11]

The Contribution of the Environment

I have discussed the division of linguistic labor and the special role played by experts of different sorts, such as people who know how to identify gold and people who know how to tell a beech from an elm. But there is a factor that I have so far neglected—an all-important one. This is the role of the environment itself (of the things we are referring to themselves). This is, perhaps, easiest to see in the case of substance terms, such as "gold" and "water." In an earlier publication[12] I illustrated the way in which the reference of the term "water" is partly fixed by the substance itself with the aid of a thought experiment involving "Twin Earth." We imagine that the year is 1750 (both on Earth and on Twin Earth) and that Daltonian chemistry has not yet been invented. We also imagine that the people on Twin Earth have brains identical with ours, a society virtually identical with ours, and so on. In fact, the only relevant difference between Earth and Twin Earth in my thought experiment was that the liquid that plays the role of water on Twin Earth was supposed not to be H_2O but a different compound, call it XYZ. On Twin Earth it does not rain H_2O but it rains XYZ, people drink XYZ, the lakes and rivers are full of XYZ, and so on. The claim I made in "The Meaning of 'Meaning'"—a claim which has provoked a great deal of subsequent discussion—is that one should say, imagining this case to be actual, that the term "water" did not have the same *reference* (even in 1750) in Earth English and in Twin Earth English. The reference of the word "water" on Earth, according to me, was the stuff *we* call water, the stuff we have discovered to be H_2O. The stuff that they called "water" on Twin Earth in 1750 (and call "water" now as well) is the stuff that fills the lakes and rivers on Twin Earth, the stuff that they later discovered (when they developed sophisticated chemistry) to be XYZ. Not only does the word "water" have a different reference *now*—now that we know that "water is H_2O" and they know that "water is XYZ"—it had (according to me) a different reference *then*.

Why do I say this? Here it is useful to recall a number of things about the way in which we view water (and, to some extent, substances generally).[13] In ancient and medieval times, water was thought of as a pure substance (in fact, it was thought of as an element by many of the ancient thinkers and by most of the medievals). Part of the notion of a pure substance is that any bit of it is expected to exhibit the same behavior as any other bit of it. People two thousand years ago, people in 1750, and people now after the rise of modern chemistry, all expected any sample of pure water to behave the same way as any other sample of pure water. If you had asked a person living in 1750 the hypothetical question, "Suppose that I gave you a glass containing 50 percent normal water and 50 percent some substance which is not found as a constituent of normal water, but you couldn't tell this by the appearance or taste or aftereffects, or by washing clothes in it, or anything like that (apart from using a still); would that mixture then simply *be* water?" I think that even in 1750 a typical person would have answered, "No, I wouldn't say it was water, I would say it was a mixture of water and something else." Of course, if it had turned out that normal water was itself a mixture, and that it contained an indefinite number of different "pure" constituents, then the answer might have been different. But we might say that our intention, even in 1750, was somewhat as follows: On the assumption that normal water is in fact a pure substance, then we do not intend the description "water" to be true *tout court* of anything which consists to a significant extent (say, 20 percent or more) of any other substance.

Now, Earth water and Twin Earth "water" were different substances even in 1750; it's just that no one on Earth or Twin Earth had yet noticed this—in fact, they didn't even know of the existence of the other substance in each case. Someone on Earth in 1750, if he had been taken to Twin Earth on a spaceship by a more advanced civilization, would have *taken* Twin Earth water for water, but he would have been making a mistake; he would have been thinking that it was the *same* substance that he knew by the name "water" on Earth. Similarly, someone from Twin Earth would have been mistaken in thinking that Earth water was what his community called "water." No one on Earth or on Twin Earth would have noticed that the word had a different meaning in 1750, but in my view they *would* have had a different meaning. The "mental representations" of Earth speakers and Twin Earth speakers were not in any way different; we may suppose that they were exactly the same, even if we include "mental representations" in the heads of the chemists; the reference was different because the *substances* were different. This illustrates how the reference is partly fixed by the environment itself. This is the phenomenon that I have called *the contribution of the environment*.

In "The Meaning of 'Meaning'" I expressed this by saying that even in 1750 what the word "water" referred to in Earth English was H$_2$O (give or take impurities). The word "water" in Twin Earth English referred to XYZ (give or take impurities). To say that this is what the word "water" referred to in the two dialects is just to say that this is what the word *denoted* or was true of; it is not to say that Earth speakers in 1750 *knew* that the word "water" referred to H$_2$O or that Twin Earth speakers knew that their word "water" referred to XYZ. But then, some have objected, it seems that I am saying that we "didn't know the meaning of the word 'water'" until we developed modern chemistry.

This objection simply involves an equivocation on the phrase "know the meaning." To know the meaning of a word may mean (*a*) to know how to translate it, or (*b*) to know what it refers to, in the sense of having the ability to state explicitly what the denotation is (other than by using the word itself), or (*c*) to have tacit knowledge of its meaning, in the sense of being able to use the word in discourse. The only sense in which the average speaker of the language "knows the meaning" of most of his words is (*c*). In that sense, it was true in 1750 that Earth English speakers knew the meaning of the word "water" and it was true in 1750 that Twin Earth English speakers knew the meaning of their word "water." "Knowing the meaning" in this sense isn't literally knowing some *fact*.

Another objection that I have sometimes encountered to my Twin Earth example is the following: people have supposed that if XYZ plays the role of water on Twin Earth, then it must exhibit *exactly* the same behavior as water on Earth, at least at the "observable" level. But this is simply a mistake. The average English speaker in 1750 was aware of only a very limited range of observable properties of water. Even the chemists in 1750 were aware of only a limited range of properties. They knew, for example, the boiling point of water (although not with present-day accuracy). They knew the density of water. They certainly did not know *all* of the chemical reactions into which water enters. However, H$_2$O and XYZ are supposed to be different compounds. Thus, there has to be some third substance S such that

H_2O chemically reacts with S in one way (perhaps in the presence of catalyst C, or in the presence of heat, etc.) and XYZ reacts with S in a different way (perhaps in the presence of catalyst C, or in the presence of heat, etc.). For example, it may be that when water is mixed with S and C is added and the mixture is heated, then the mixture turns green and drops a yellow precipitate, whereas when Twin Earth water is mixed with S and C is added and the mixture is heated, then one gets a tremendous explosion. (Or it might simply be that Twin Earth water fails to react with S at all, or reacts only with a different catalyst.) This phenomenon (and many other similar ones) would show that Earth water and Twin Earth water are two different substances. But that does not mean that the "mental representations" were different in 1750, because neither Earth speakers of English nor Twin Earth speakers of English knew of these facts back in 1750. In short, the "mental representations" were the same in 1750, and yet the reference was different, and moreover this difference in reference *could* have been shown to Earth people and to Twin Earth people who were alive in 1750 notwithstanding the "sameness of their mental representations."

An Indexical Component

What makes this possible is what I have called the *indexicality* of our criteria for being water (for being a sample of a particular substance). There is a "property" which people have long associated with pure water and which distinguishes it from Twin Earth water, and that is the property of *behaving like any other sample of pure water from* our *environment*. To use a term suggested by Alan Berger,[14] when we teach the meaning of the word "water," we *focus* on certain samples. A substance which doesn't behave as *these* examples do will be counted as not the same substance (barring a special explanation). But the "property" of "behaving the way *this* stuff does" isn't what philosophers call a purely "qualitative" property. Its description involves a particular example—one given by pointing, or by "focusing" on something. Now, if the water I am focusing on looks and tastes just like the "water" that Twin Earth Hilary is focusing on, then my "mental representation" of my example may be "qualitatively" identical with Twin Earth Hilary's representation of his example. But the *stuff* is different, and so the *property* of being-pretty-much-like-*this* is a different property when I define it that way from the property of being-pretty-much-like-*this* which Twin Earth Hilary defines that way. Property terms of this kind, property terms which contain words like "this" or "here" or "now," can refer to different properties in different circumstances of use. In short, we *had* a criterion in 1750 which distinguished Earth water from Twin Earth "water"; it was not a qualitative criterion, but an indexical criterion.[15] That indexical criterion was associated with exactly the same mental representation that Twin Earth speakers of English would have used (had Twin Earth actually existed) to distinguish Twin Earth water from Earth "water." It is because the two different criteria were both indexical that they could be associated with *identical* mental representations in the heads of speakers in the two different communities and still pick out *different* substances (just as the mental representation *the conductor of this bus* may be ever so identical in quality in two different heads and still pick out different individuals).

Other Natural Kinds

Once the point is grasped in the case of substances, it can easily be extended to other natural kinds. Using the fiction of Twin Earth once again to illustrate the point, it could be that the mental representation associated with "cat" on Twin Earth in 1750 was exactly the same as the mental representation associated with that word on Earth in 1750 although Twin Earth cats are a totally different biological species (have different DNA, are not cross-fertile with Earth cats, and so on). What our discussion shows is that an ideal interpreter could not know whether the Earth term "water" and the Twin Earth Term "water" have the same meaning (should be translated in the same way) without knowing a certain amount about both Earth and Twin Earth chemistry; that he could not know whether the Earth term "cat" and the Twin Earth term "cat" have the same meaning without knowing a certain amount about both Earth and Twin Earth biology; and so on.

In certain ways, the case of biological species is different from the case of pure substances, however. Pure substances are a somewhat special case. The belief that any sample of a pure substance will exhibit the same behavior as any other sample of the same substance is only one of the beliefs which help us to fix the reference of terms which refer to such substances. Another, equally ancient, belief is that any two such samples have the same ultimate constitution. (I don't think, however, that this is really a totally different criterion from the "same behavior" criterion. For we expect differences in ultimate constitution to show up as differences in behavior and differences in behavior to be "grounded" in differences in ultimate constitution.)[16] Thus the fact that Twin Earth water was *not* water, not even by the standards of 1750, even if one would have had difficulty finding a way of proving this in 1750 (unless one were a genius), is overdetermined. Twin Earth water violates (and always violated) two conditions for being called "real" water: it neither has the same ultimate constitution as "our" water nor exhibits exactly the same behavior.

I have dwelt at length on this case because I think it is in certain ways simpler than the other cases, but I think that similar principles apply to other natural kinds. We do not expect any two members of a biological species to exhibit the same behavior or to have exactly the same appearance (Siamese cats do not have exactly the same appearance as European cats); but we do have the expectation that (with occasional exceptions) two members of a species who are of opposite sex and who are biologically fertile will be able to mate and to have fertile offspring. If Twin Earth "cats" were never able to mate with Earth cats (and produce fertile offspring), then not only biologists but laymen would say that Twin Earth cats are another species. They might, of course, say that they were another species of *cat*; but if it turned out that Twin Earth cats evolved from, say, pandas rather than felines, then in the end we would say that they were not really cats at all, and Twin Earthers would similarly say that Earth cats were not really cats at all.

Moreover, in the case of what look to be biological species, questions of ultimate constitution may also enter. If we suppose that Twin Earth cats look exactly like Earth cats and behave exactly like Earth cats, but it turns out upon detailed scientific examination using sophisticated theory and technology that they are really robots remotely controlled from Sirius, not only will we say that they are not really cats (in the Earth sense), but we will say that they are not really *animals* in the Earth sense at all. Whether they are "animals" in the Twin Earth sense will depend on whether the Twin Earth dogs, lions, tigers, etc., are or are not also remotely controlled from Sirius. If all "animals" (except people) on Twin Earth turn out to be robots remotely controlled from Sirius, then a Twin Earthian might well say, "That's what animals *are*," whereas an Earthian will say, "They aren't really animals."

Still another case is that of highly impure kinds, such as mixtures of one sort or another. We do not expect any two samples of *milk* to exhibit exactly the same behavior—some milk has higher butterfat content, some milk tastes of clover while other milk does not, and so on. There may even be a small percentage of constituents in some milk that do not occur in some other milk. But if something does not consist at least 50 percent of the constituents that we find in "normal" milk, then even if it tastes like milk, we will say that it is not "really milk" (although we might say that it "contains milk"). The point of all these examples is the same. The description given by both the Earthians and the Twin Earthians of X, where X is gold, or cats, or water, or milk, or whatever, may be the same (apart from the difference in the reference of the indexicals "we," "here," "this," etc.); the mental representations may be qualitatively the same; the description given by the experts at a given stage of scientific development may be the same; but it may turn out, because of the difference between the Earth and Twin Earth environments, that the *referents* are so different that Earth speakers would not regard the Twin Earth gold as gold at all, or regard the Twin Earth water as water at all, or regard the Twin Earth cats as cats at all, etc. *Meaning is interactional. The environment itself plays a role in determining what a speaker's words, or a community's words, refer to.*

Reference and Theory Change

We must now take a closer look at how reference is fixed. In the last section I said that the reference of a word like "gold" is fixed by criteria known to experts, and that it doesn't matter if the experts use different criteria, as long as the same stuff (apart from borderline cases) passes the various tests that these experts use. This is compatible, as I said, with East Coast experts using different tests from West Coast experts, or American experts using different tests from Asian experts, or experts in the twentieth century using different tests from experts in the nineteenth century, etc. If this were all there is to say about the fixation of reference, then when the different tests do not exactly agree, the cases on which they disagree would be correctly classified as vague cases, or ambiguous cases, or something of that kind. We could then say that something is gold if it passes *all* the tests used by experts in all the centuries and all the places; that it is not gold if it fails all the tests used by experts in any century and any place; and that it is a "borderline case" if it fails some tests and passes others. But in view of what I have just said, we can see that this would be wrong. For it makes sense to say that some of the tests are not *correct*.

If the tests for gold in use prior to Archimedes could have been passed by some stuff that did not in fact have the same density as gold, then those tests were incorrect, and Archimedes found a way of showing that those tests were incorrect and of correcting their results. He did this by relying on the principle I mentioned, that any sample of pure gold should exhibit the same behavior as any other sample of pure gold. By finding out how to determine the density of metals, he found a way to investigate the behavior of samples with respect to a parameter no one previous to him knew how to measure. Now that we have developed ways of determining the atomic constitution of a substance, and even the constitution at the subatomic level, we have still better means of determining when and how our tests fail. If people at some previous time woud have accepted some alloy as gold, that does not mean that it was gold, in the sense in which the word "gold," or "chrysos," or whatever, was used at that time; it means that the people at that time did not have a way of *knowing* that they were dealing with something that had neither exactly the same behavior nor exactly the same constitution as the paradigm examples of gold. They did not know that the alloy was not really gold. But what they meant by "gold" (or "chrysos," etc.) was what we mean by "gold." The fact is that no set of operational criteria can totally fix the meaning of the word "gold"; for as we develop better theories of the constitution of gold and more elaborate tests for the behavior of substances (including the behavior in respects that we were not previously able to measure), we can always discover defects in the tests that we had before.

The same thing goes for natural kinds which are not substances. Suppose the Martians are able to build robots that look exactly like animals—they even have organic bodies, and their brains are full of stuff that looks to present-day scientists exactly like brain matter (although it doesn't really *function* as such)—but these "animals" are really directed by signals received by a miniaturized radio receiver implanted in the pineal gland. (The "brain" is just an elaborate fake.) Suppose a few of these have been smuggled in among the "normal" animal population, but most animals are the naturally evolved organisms we take them to be. Then when we develop the scientific resources to detect the fakes, we shall say that the fakes in question are not really animals (and not really cats, or whatever); even though up to this point, they may have passed all of our operational tests for being animals (and for being cats, etc.). Thus the fact that the environment itself contributes to the fixing of reference is also one of the reasons that naive operationalism and naive verificationism are wrong as an account of the meaning of natural-kind terms.

Meaning and "Mental Representation"

So far I have suggested that traditional mentalistic accounts of meaning and reference fail in two different ways. On the one hand, they neglect the division of linguistic labor. On the other hand, they neglect the way in which the paradigms that are supplied by our environment contribute to the fixing of reference. Because of these oversights, the traditional theorist is unable to imagine how two speakers or two communities could associate the very same "mental representations" with terms and yet use the terms to refer to different species, substances, etc.

This does not mean that descriptions, including descriptions "in our minds," play no role in fixing reference. Both nonindexical descriptions (the descriptions of the behavior and/or composition of gold that an expert might give) and indexical descriptions ("stuff that behaves like and has the same composition as *this*," said by someone who is "focusing" on a particular sample of a substance) do help to fix the reference of our terms. Indexical descriptions can be extremely important in fixing reference, but, as we have seen, they are not what we preserve in translation. The term "gold" is not *synonymous* with "stuff that passes the following test," or with "stuff that has the same behavior and ultimate composition as this." In fact, the effect of my account, as of Saul Kripke's in *Naming and Necessity*,[17] is to separate the question of how the reference of such terms is fixed from the question of their conceptual content.[18]

In the face of the difficulties I have been describing, some authors, reluctant to give up the whole of the Aristotelian picture, have tried to see if they could retain at least two of the three assumptions. I have already mentioned the case of John Searle, who has indicated that he, at least, is prepared to give up assumption 2 in my list of Aristotelian assumptions (this is the assumption that sameness of mental representation just *is* sameness of meaning, or "synonymy"), in order to hang on to the other assumptions. I have also mentioned in passing that Fodor (and at times Chomsky) would hold on to 1 and 2 while giving up 3 (the assumption that mental representation is what fixes reference).[19] Before we examine this suggestion, it will be useful to take a closer look at a notion I have so far been employing uncritically (the way, I fear, many psychologists now employ it), the notion of a *mental representation*.

At a surface level mental representations do not differ very much from representations by means of spoken sounds, or by means of writing, or from other signs. Just as one can write the words, "There are a lot of cats in the neighborhood," so one can say them, one can store them on a floppy disk, and one can also think them without speaking out loud. The notion that unspoken thought is simply subvocalization may be an extreme bit of reductionism, but it has a point. There is not much difference between *words* in one medium—even a mental medium—and words in another, just as words. However it is that words like "cat" and "neighborhood" manage to refer, it is not just by having a certain spelling or a certain sound—not even a certain spelling or sound "in one's mind." These surface representations—spoken thoughts—cannot be the concepts that Aristotle referred to, nor are they the "mental representations" that modern mentalists are talking about. The representations spoken of in 1, 2, and 3 were supposed to be representations which *determine* the meaning of words, not words themselves, and they were supposed to be the same whether one uses the word "elm" or the word "orme" to refer to elms.

Distinguishing between surface mental representations ("subvocalized" thoughts) and deep mental representations does not affect our criticisms of the Aristotelian Theory, because if someone is totally ignorant of the differences between an elm and a beech (he only knows that there *are* differences), then this ignorance must extend all the way down; we cannot suppose that although his surface representations do not distinguish between elms and beeches, his "deep" representations somehow do, for he has never *learned* the difference. No matter what we postulate in the way of "deep," or "underlying," or "unconscious" mental representations, we can reasonably suppose that at every level, no matter how deep, *my* mental representation of an elm is identical with my mental representation of a beech, except as concerns my knowledge of the different phonetic shapes of the names "elm" and "beech"; and similarly, we can suppose that my mental representation of a beech could be the same at every level as a Frenchman's mental representation of an "orme," apart from phonetic properties.

The problem with mental representations at the level of conscious thought—which are the only mental representations of whose existence we have any sure knowledge—is that they badly violate principle 2. The Frenchman's surface mental representation of an elm is not literally the same as my surface mental representation of an elm. His mental representation, at the surface level, is *arbre qu'on appelle "orme"*; my mental representation, at the surface level, is *tree that one calls an "elm."* These are not literally (syntactically) the same object. We could decide in certain contexts to treat them as the same: we might just decide to *identify* mental representations that are synonymous. Such a maneuver would buy us nothing. The idea that what synonymy *is* is being associated with the same mental representation assumed that we had a notion of *identity of mental representation* independent of the notion of synonymy. If the very notion of having the *identical* mental representation is really just a *façon de parler* for "having mental representations with the same meaning," then assumption 2—the assumption that synonymous expressions are associated with identical mental representations—becomes trivially true. (Expressions with the same meaning are, among other things, associated with *themselves*, and they themselves are mental representations with the same meaning.) It is for this reason that Fodor has to postulate a *lingua mentis*, often called "Mentalese" in his writing, and a Cryptographic Model of the mind, according to which when a Frenchman thinks (at the surface level), *Les ormes sont arbres*, this gets transcribed into a formula or sentence in Mentalese which is exactly the same—identical by a *syntactic* criterion of identity[20]—as the formula in Mentalese which the Cryptographer in my brain encodes in English as *elms are trees*. If Fodor's theory is right, Aristotelian assumption 2 is correct, and assumption 2 is no tautology.

Assumption 2 is not a tautology, in Fodor's theory, precisely because the identity (or equivalence) relation between mental representations in Mentalese is supposed to be defined *syntactically*.

What of assumption 3? We have just seen that even if Fodor's theory is correct, it cannot be supposed that identity of "mental representation" always guarantees identity of referent (e.g., the elm/beech case, as well as the case of Earth water and Twin Earth water). Fodor concedes this point. His response in a number of papers[21] is to say that the ordinary notion of meaning is referentially *ambiguous*.[22] One referent ("narrow content") is mental representation at the deepest level (the "semantic representation" in "Mentalese"). Another referent ("broad content") is the function which gives the referent(s) in each possible world.

The notion of "broad content" evidently depends on the notion of *reference*. This notion (reference) Fodor hopes to explicate with the aid of the notion of *causality*. Projects of this kind—attempts to explicate the notion of reference—will occupy our attention in the remaining chapters of this work. For this reason, I shall not discuss the notion of "broad content" further now.

Given that Fodor does not intend his work as a conceptual analysis of the notion of *meaning*, but rather as an empirical theory about the workings of the human mind, it might appear puzzling at first blush that he thinks that the "Mentalese" hypothesis has anything to do with our topic. If his theory claimed that "mental representations" somehow fixed reference, then if his theory were scientifically spelled out and scientifically verified, it would constitute a vindication of the entire Aristotelian view. But by separating "narrow content" from "broad content" and admitting that the "narrow content" of a term does not determine its "broad content" (for just the reasons we have given above), Fodor blocks this defense of the philosophical significance of his theory.

Suppose the theory is right; then, when the Frenchman thinks (in French), *Il y'a beaucoup des ormes dans le voisinage*, he thinks a sentence which encodes a formula in Mentalese, as it might be "φ©ηΔΔιʻ." When I think (in English), *There are a lot of elms in the neighborhood*, this is simply the way my brain encodes the same formula, "φ©ηΔΔιʻ" (or an "equivalent" formula, under some *syntactically* definable equivalence relation). To take a simpler example, when I think the word *cat*, then, according to Fodor's theory, the Cryptographer in my brain "decodes" this as, say, "*#@å," and when a Thai speaker thinks the word *meew*, this is simply the code used by the Cryptographer in *his* brain for "*#@å." This is fascinating if true, and a contribution to our understanding of the way the brain works (if true), and, perhaps, very important in psychology (if true), but what is its relevance to a discussion of the *meaning* of *cat, meew*, or "*#@å"?

Chapter 3
Fodor and Block on "Narrow Content"

Different answers have been suggested to the question with which I ended the preceding chapter. Fodor's current view[1] is that the narrow content is not the mental representation at all (as suggested in his earlier writing, including *The Language of Thought*); instead, narrow content is a "function" from context to referent. (How this is supposed to be in the speaker's head is something I don't understand; in addition, as I have already remarked,[2] this theory is, as far as I understand it, open to all the objections I made to Searle's theory.) In a series of papers which Fodor wrote in response to my "Meaning Holism,"[3] but did not publish (because he changed his mind), Fodor proposed an interesting and rather different view, which I think is worth discussing even if Fodor himself has abandoned it.

Narrow Content as a "Function of Observable Properties"

Fodor's proposal was that words (in "Mentalese") are semantically associated with perceptual prototypes (and thus with "functions of the observable properties" of various things). We have a perceptual prototype of the dog, and we often recognize that an animal we meet conforms to that perceptual prototype before we recognize what particular sort of dog it is,[4] terrier, retriever, collie, or whatever. Indeed, this fact, the fact that we have a stereotype of a dog or a typewriter or a table or a tree, has long been noticed by philosophers and psychologists. In the days of Berkeley and Hume, such prototypes were referred to as "ideas," and there was considerable discussion about their nature. Some seventeenth-century psychologists thought that they are simply mental images; but although the study of mental images continues to be an active topic today,[5] there are well-known difficulties with this view. My perceptual prototype of a dress, for example, may include a certain shape, but it does not include any particular color. Yet any particular image of a thing, e.g., a dress, must have some particular color. In short, my perceptual prototype of a dress is more *abstract* than any one mental image. (A problem that was much discussed in the seventeenth century.) Another approach to perceptual prototypes (or stereotypes) simply identifies them with verbal responses. Thus, if upon being asked to describe a "typical tree," I reply that I would expect a typical tree to be at least fifteen feet tall and to shed its leaves in winter, then *being at least fifteen feet tall* and *being deciduous* would be counted as parts of my stereotype of a tree. This notion of a stereotype is of no use for Fodor's purposes, because stereotypes (in this sense) are words and phrases, and Fodor's whole purpose is to get to something which lies *behind*, something which is deeper than, words. However, Fodor believes that one can escape the trap question, "Are perceptual prototypes themselves verbal, or are they images?" by proposing a third alternative, derived from modern computer science.

It is plausible, in terms of present-day brain research, work on artificial intelligence, and so on, that the brain contains devices for recognizing *patterns* (or more generally, "functions of observable properties"). Such devices are one kind of *module*, in Fodor's sense. When I have learned to recognize dogs, then, Fodor postulates, somewhere in my visual system I have built up a subroutine in a little computer, one which functions independently of general intelligence, which has the function of recognizing things that exhibit particular patterns. It is quite possible for a pattern-recognition device to recognize shape without recognizing color. Thus the difficulty just raised against the identification of perceptual prototypes with images does not bar the identification of perceptual prototypes with the outputs of particular subroutines executed by modules.

Moreover, a module (or a subroutine) does not, by itself, have anything like the "understanding" of a complete language. Just as the thermostat in my furnace can "recognize" changes in temperature without possessing the concept of temperature, so a pattern-recognition device in my visual system might be able to recognize anything with the shape of a dog without possessing the concept of a dog.

That something like this must be possible is indicated by Herrnstein's work with pigeons. Herrnstein has shown that pigeons can be trained to do extremely sophisticated recognition of different types of objects, not only from direct perception, but even from photographs; yet the pigeon's brain is minimal compared to the human brain. The pigeon that recognizes a building in a blurred photograph is not going through any reasoning in a natural language; presumably it is exercising modularized pattern-recognition routines.

We can now restate Fodor's view of narrow content in a way that at least partly meets the objection that, so far, narrow contents are just uninterpreted formulas in a hypothetical language called Mentalese. The formulas in the brain's system of representation, or "language," do have an association with something nonlinguistic, even if in most cases that something does not determine the reference of our terms. The words "elm" and "beech" are associated with the stereotype of a deciduous tree, for example. Speaking the language of Fodor's theory, one might say that the words "elm," "beech," and "common deciduous tree," for example, all have the same "narrow content," and that the sense in which they differ in meaning is that they have different "broad content." The narrow content of the words "elm" and "beech" (the associated "pattern" in the appropriate visual module) does not enable me to recognize elms and beeches. Nevertheless, it does contain substantive empirical information about the observable properties of elms and beeches—viz., that they are common deciduous trees.[6] In Fodor's view (and also, as I know from discussions with him, in Chomsky's), what the division of linguistic labor shows is that the narrow content of a term may contain some information about the observable properties of a referent without containing enough information of that sort to enable me to actually *identify* the referent. For example, if I cannot tell silver from whatever alloy the United States government is currently making its dimes and quarters out of, that does not show that "silver" and "silvery metal" have the same *meaning*, in the sense of the same broad content, but it does show that for me "silver" and "silvery metal" have the same narrow content. By associating "semantic representations" in the fundamental sense of "formulas in Mentalese" with modules (or, more precisely, with particular outputs of modules, or particular decisions by modules),[7] Fodor attempted to give the theory of narrow content psychological substance. In the fashion of many mentalistic theories, he did this by pairing linguistic mental entities with nonlinguistic (but still mental) entities which carry a "content."

There are, however, serious difficulties with this theory that must be pointed out. First of all, it is essential to Fodor's basic claim (the systematic referential ambiguity of the notion of *meaning*) that *every* term in the language has a narrow content, and not only terms which correspond to perceptual prototypes. Indeed, Fodor has argued that it is impossible to give an account of *belief* unless everything that can be believed has narrow content.[8] But the word "zeitgeist," for example, is not associated with a perceptual prototype, although one can certainly have beliefs about the zeitgeist, e.g., that *the rabid individualism we now see is an expression of the zeitgeist*. That narrow content is a function of the *observable properties* of the referent or supposed referent is at best a *fragment* of a theory of narrow content.

Even as a fragment of a theory (say, as a theory for terms referring to "animals, vegetables, minerals, and middle-sized dry goods"), there are serious difficulties with the idea that "narrow content is a function of the observable characteristics." These difficulties are of two sorts: perceptual prototype is not preserved in translation, and, in some cases, it is not what is important for translation.

To see that perceptual prototype is not preserved in translation, recall that all the cats a typical villager will have seen in Thailand are Siamese cats. His perceptual prototype of a *meew* is our perceptual prototype of a "Siamese cat," *not* our perceptual prototype of an "ordinary cat." Yet the correct translation of *meew* is "cat," and not "Siamese cat" (i.e., *meew* denotes all cats, not just ones which resemble the Thai stereotype). (Think also of the perceptual prototype of a dog—someone who has seen only huskies will have a different prototype of a dog from someone who has only seen dogs in Mexico,[9] yet it will be correct to translate his word for the species as "dog.")

To see that perceptual prototype may not even be important for translation, imagine a culture which has the traditional notion of a *witch* (i.e., of a woman who has magical powers, usually exercised for evil). Suppose the perceptual prototype associated with the word is "ugly old woman with a big nose and warts." Still, the *meaning* of the word "witch" in this culture is not in any sense correctly rendered as "ugly old woman with a big nose and warts"; that may be the stereotype, but there is conceptual content to the word which is more important than the stereotype, viz., the imputed magical powers. Perceptual prototypes may be psychologically important, but they just aren't *meanings*—not even "narrow" ones.

"Narrow Content" and "Conceptual Role"

I mentioned earlier that there is an alternative account of narrow content due to Ned Block.[10] Block's theory resembles Fodor's in accepting the distinction between "narrow content" and "broad content." Also as in Fodor's theory, "narrow content" is supposed to be a mentalistic notion—narrow contents depend only on what is inside the speaker's head, and can be described without taking into account the extension of a term or sentence. (So that "water" has the same narrow content on Earth and on Twin Earth but not the same broad content, as in Fodor's theory.) Here the resemblance to Fodor's theory ends; although Block agrees with the dominant computational approach in thinking of the mind/brain as computing with "representations" in a "language of thought," he does not take a stand on whether this is an innate language (Fodor's Mentalese) or simply the local natural language (suitably transcribed into neuronal terms), nor does he re-gard narrow content as a sense of "meaning." (He does, however, sometimes use the expressions "broad meaning" and "narrow mean-ing," and also speaks of "two ways of individuating thought con-tents"—just as Fodor does.) Rather than regarding the notion of "meaning" as ambiguous and taking "narrow content" and "broad content" to be the two senses (or rational reconstructions of the two senses), Block regards meaning as, in effect, the ordered pair of nar-row content and broad content. For Block, narrow content is one of the two factors *determining* meaning, rather than a *kind* of meaning (or at least this is his fallback position).

Moreover, narrow content is not identified with a function of ob-servable properties, as it was in Fodor's (former) theory. Rather, it is identified with "conceptual role."

What is conceptual role? Block gives a number of people[11] credit (he even cites Wittgenstein's remark that for many purposes we can think of meaning as use as representing this approach). Perhaps the clearest origin of the notion is in the work of Wilfrid Sellars,[12] who describes language as having "language-entry rules" (think of these as rules saying that when the speaker has certain experiences, he is to put certain sentences in the "belief box"), "language-language" rules (rules that when the speaker accepts certain sentences, he is to accept certain other sentences), and "language-exit rules" (rules say-ing that when the speaker has certain sentences in the belief and desire boxes, he is to perform certain bodily movements, or say cer-tain words, etc.). (The "language-language" rules are not all required to be analytic—there can be material rules of inference, according to Sellars.) The important thing (in the Sellars-Block conception) is that (apart from the reference to experiences—Block would no doubt re-place this by a reference to outputs from Fodor's pattern-recognition modules—in the case of the language-entry rules, and to bodily movements in the case of the language-exit rules) the conceptual role of the formulas of a language can be described entirely *syntactically.* If we had the computational description of the syntactic processes into which words and sentences enter, then we would be in a position to computationally define the relation of similarity of narrow content: words are similar in narrow content if they have similar "conceptual roles" (in the Sellars-Block sense) in their respective languages. And if words are similar in narrow content and also have the same refer-ence (or better, the same broad content), then they are similar in *meaning*.

Block speaks in terms of similarity of meaning rather than in terms of sameness of meaning because he believes that we shall have to "do away with the crude dichotomy of same/different meaning in favor of a multidimensional gradient of similarity of meaning."

There are a number of points that I want to make in connection with the claim that the component of meaning that is responsible for making meaning more finely individuated than it would be if meaning were simply *identical* with broad content[13] is "conceptual role." If we take conceptual role to be a matter of what beliefs containing a word are most central to the topic that word picks out, what inferences are similarly central, what practices the word is connected with, and so on, we have already seen that conceptual role in this sense has little to do with meaning in the case of the important class of words on which I have so far concentrated: the natural-kind terms. In the case of the English word "water" and its cognates (e.g., the Greek "hydor"), there has been an enormous change in the conceptual role through the centuries, and yet we do not regard these words as having changed their meaning (or at least most of us don't). In ancient Greece, "hydor" was not only the substance that we drink, but it was also the name of an element, and as an element it was virtually a universal principle of liquidity. This way of conceiving of water survived into the Renaissance, and its traces survive in such expressions as "aqua vite." That *alcohol is liquid because it contains the element water* and that *mercury is liquid because it contains the element water* are propositions which seem decidedly odd today, but they would have seemed to be true statements in ancient Greek times and, as I just remarked, even to many Renaissance thinkers. (There was also a conception of matter intermediate to the classical four-elements theory and the Daltonian theory—this was the theory, widely accepted by physicists in Newton's time, that matter consists of atoms, but those atoms are the same in all substances: gold is not different from lead because the atoms are different, but because the arrangement is different. It is because he accepted this conception that so great a physicist as Boyle held that any substance can, in principle, be transformed into any other. Boyle would not have seen any sense to saying that ice is "water in a frozen state." According to him, ice and water are as different as water and lead; freezing water is *transmuting it into* ice.)

One conclusion—the one I myself would draw—would be that overall conceptual role can change enormously without there being a change in meaning; that is, one must abandon Conceptual Role Semantics. Or one might try to get around this by saying that when someone says, "May I give you a glass of water?" or when someone said the corresponding words in classical Greek, he was using the word "water" or "hydor" in an "ordinary" sense, and when he said that all liquids have water in them he was using the word in a different sense (a "philosophical" sense). But, as I argued above, this move fails to do justice to the interdependence of our ordinary and our scientific uses of words. Even in its so-called "ordinary" sense, the word "hydor" has undergone enormous changes in its conceptual role since ancient times (and the cognate word "water" has undergone equally large changes in its conceptual role in English even since the seventeenth century).

Since reference is only one factor in determining meaning, on Block's theory, and the other factor has changed enormously, it would follow that the word "water" has changed meaning in this period (and it would follow that the word "water" has a different meaning than the ancient word "hydor"). Moreover, the meaning change cannot be small if the conceptual role change is large; for the whole reason for assigning weight to conceptual role in Block's theory is to explicate the sense in which there can be substantial differences in meaning in the case of words which do not differ with respect to the other factor (words which do not differ in "broad content").

At this point, however, a threat looms. If we are going to have to say that a word has changed its *meaning* whenever our beliefs change significantly with respect to the topic in question, then our use of the word "meaning" will no longer have any contact at all with the ordinary notion. Of course, it is not unreasonable to say that a notion of meaning which is constructed to be scientifically useful may differ in some cases or to some extent from the ordinary notion of meaning. For example, I have already mentioned Block's suggestion that we give up the "crude dichotomy of same/different meaning" in favor of a "multidimensional gradient" of similarity of meaning. But if words that are treated as having the same meaning in actual translation practice do not turn out even to have *similar* meanings, after the notion of meaning has been reconstructed, then it would seem that the reconstruction is really just a change of subject, and not a theory of meaning at all.

The cause of the problem is not hard to see. The fact is that when a word is a natural-kind word, we generally translate it by the corresponding natural-kind word in our own language, where the corresponding natural-kind word is the natural-kind word that has the same *extension*. There are exceptions to this which I shall discuss, but in the case of natural-kind words it seems that the dominant "component" of meaning is the extension. The referential factor seems to do almost all the work; in this respect, as Saul Kripke pointed out, natural-kind terms resemble *names*.

This does not mean that the natural-kind term "water" is synonymous with "H$_2$O," or even with "H$_2$O give or take some impurities"; for although "water" functions as what I have been calling a "natural-kind term," "H$_2$O" functions quite differently. "H$_2$O" is synonymous with a description, namely, the description "chemical compound which consists of two parts hydrogen to one part oxygen"; and I am not saying that an expression which has the logic of a natural-kind term has the same meaning as an expression of a different kind (a description) which has the same extension.

What is it to be a natural-kind term, however? Block might claim that to be a natural-kind term is just to have a certain (rather difficult to describe) conceptual role. This is perhaps true. (Although it is not clear that that role can be described without using semantical terms— for example, that a natural-kind term is not synonymous with the descriptions we use to fix its reference is in some sense a statement about the "conceptual role" of natural-kind terms, but *this* description uses the semantical term "synonymous.") It may be—I grant this for the sake of argument—that it is the conceptual role—in Block's sense of "conceptual role"—that a word plays that identifies it as a natural-kind term; however, I claim that once we have identified a word as a natural-kind term, then we determine whether it is synonymous with another natural-kind term primarily on the basis of the extensions of the two words. In short, in the case of natural-kind words the conceptual-role component plays only a limited role in determining the meaning, and most of the work is done by the second factor. There is no way around the fact that natural-kind terms are an exception to Block's theory. There are many other sorts of words, however, and it might be supposed that the theory fares better in other cases—that names (and natural-kind terms are often classed with names, nowadays) are an exception to almost any semantic theory. So let us look at words other than natural-kind words.

Before we do this, I should mention an exception to the claim just made. Some words which were intended to be natural-kind terms turn out not to refer to natural kinds. "Phlogiston" was intended to be the name of a natural kind, but it turned out that there was no such natural kind. And similarly for "ether" and "caloric." In these cases it does seem that something like conceptual role is the dominant factor in meaning, for obvious reasons; we don't want to say that the words "ether" and "caloric" and "phlogiston" are synonymous because they have the same (empty) extension. Not having an extension[14] (that is, lacking a *nonempty* extension) to constitute the, so to speak, individuality of the word, one naturally falls back on the conceptual role. Indeed, the conceptual role theory comes closest to being true in the case of words with an empty extension.

Even in the case of these words, however, it is not *overall* conceptual role, as constituted by the totality of the beliefs and inferences that a speaker would regard as important or central in connection with the topic, that constitutes the "meaning" of the word. In the case of the word "witch," for example, beliefs—even "central" beliefs—have changed enormously through the centuries. The belief that seems to fix the meaning of the word has remained very stable, namely the belief that witches, if there are any, are female and have magical powers granted to them as the result of a pact with the Devil. Someone might say that this last is an "analytic" belief, and that the problem of characterizing "analyticity" is a problem for *every* theory of meaning, not just for the conceptual role theory; but this would be wrong. What I have just cited is what I called earlier a "stereotypical" belief, a belief about what a *paradigmatic* witch is like, and not at all an analytic statement. The Good Witch in *The Wizard of Oz* presumably does not owe her magical powers to a pact with the Devil; and there are witches (or words that we translate as "witch") in African languages even though there is no such figure as "the Devil" or "Lucifer" or "Satan" in the mythologies associated with those languages. A central problem that a conceptual role theory faces is this: only a *small number* of the beliefs we have at a given time partake in fixing the meaning of a term. (Moreover, it is often the case that these beliefs are not live beliefs at all, but elements of a stereotype.) The "belief" that a witch is a female human being with magical powers is central to the stereotype of a witch. That a king is the male hereditary ruler of a country is the central feature of the stereotype of a king. But in neither case do we have a necessary and sufficient condition. There are, in religious literature at any rate, female human beings who have magical powers but who are assigned to a different category than "witch," e.g., "saint," and there are kings who do not rule (the most recent king of England), and there are countries in which the kingship is not hereditary. The words "witch" and "king" are somewhat similar to the word "game," which Wittgenstein used as an example. The paradigm of a "game" is a recreational activity which requires two or more players (or teams of players) and in which players (or teams) win or lose, but the use of the word has been extended by what Wittgenstein called "family resemblance" to such activities as solitaire and Ring a Ring o' Roses. In a similar way, the word "witch" has had its denotation extended by family resemblance.

Even though they are not analytic definitions of the words "game," "witch," and "king," still certain beliefs about what a paradigm game or witch or king is like largely determine what we call the "meaning"

of these words, while other beliefs are ignored. The beliefs of an Is-
raelite two thousand years ago about the properties of a "melekh"
were undoubtedly very different from the beliefs of a typical present-
day American about the properties of a king; yet we translate "melekh" as "king," primarily because the most salient feature of the
stereotype (that a king is "the male hereditary ruler of a country")
has remained invariant. Block would reply that the nonanalyticity of
these stereotypical beliefs is just his reason for replacing "sameness
of meaning" with "similarity in meaning." "Game," he might say, is
not synonymous with "recreational activity with two or more players
(or teams) which involves winning and losing," but it is "very similar
in meaning"; and "king," he might say, is not synonymous with
"male hereditary ruler of a country," but it is very similar in meaning.
(How this could explain the fact that "In country X, the kingship is
not hereditary" is not at all contradictory, while "In country X, the
male hereditary rulership is not hereditary" is a contradiction, I shall
not even venture to guess.) And Block might say—indeed, he does
say—that any theory of meaning will have to come to terms with the
fact that certain beliefs, though not analytic, contribute more to fixing
the meanings of words than others, and we have to identify those
beliefs. All of the difficulties I have been pointing out for the Conceptual Role Theory are difficulties that must be faced by any semantic
theory, he would say, and thus they do not really count against Conceptual Role Semantics.

My aim, however, is not to advocate a different "theory of meaning." If it were, then this reply would be effective. The fact is that I
am skeptical of the whole enterprise of a "theory of meaning" in
Block's sense; that is, of a theory which is supposed to yield a *scientifically describable* relation of similarity and difference in meaning.
Against someone who doubts that such a scientifically describable
relation exists, this reply has no weight.

To recapitulate: The difficulties so far pointed out with what Block
calls CRS (Conceptual Role Semantics) are of two kinds. First, conceptual role only functions significantly to fix the meanings of certain
kinds of words, and not of others. Second, even when it does play a
significant role in determining meaning, it is not *overall* conceptual
role that does this. Words can keep their meanings invariant across
conceptual revolutions; this was precisely the discovery that moti-
vated Quine's version of meaning holism. That it is possible to pick
out *computationally* (that is, by means of predicates and relations defined in computational-syntactical terms) a set of beliefs and inferences such that those beliefs and inferences are the *core* beliefs and
inferences in the case of an arbitrary word can only be made plausible

by giving us some indication as to how we are to go about defining such a relation. If one cannot even informally indicate—without using such an expression as "regarded by speakers as part of the meaning" or "central to the meaning"—how one would decide *which* inferences and *which* beliefs fix the meaning of a word, in the sense required by CRS, then the claims made on behalf of CRS have virtually no content.

To these two points, let me add a third—and this, I think, is the most important: Let me grant, what I believe to be the case, that there is a certain insight contained in such slogans as "meaning is use" and "meaning is conceptual role." This insight, however, depends on taking the notions of "use" and "conceptual role" in a *non*mentalistic way. The meaning of the word "king" does depend on certain beliefs and inferences that people regard as stereotypical about kings.[15] But to say that, in a certain stereotype, a king is supposed to rule a country or people is not to say what the "conceptual role" of the word "king" is *in Block's technical sense*. The belief that I just described was identified *nonsyntactically;* I assumed that you understand the words "country," "people," and "rule." Conceptual Role Semantics, in Block's sense, must not identify beliefs and inferences in this way. In Block's technical sense, the "role" of a word is something that we can describe *entirely syntactically.* All we are supposed to know (apart from the connections to pattern-recognition modules and motor organs) is the syntactic properties of the mental representations inside the brain, and the syntactically-computationally described processes of inference, etc., that involve those representations. We are not supposed to know that a certain representation refers to males or to countries or to ruling when we describe the conceptual role of the word "king." The idea is to describe language as a formal system governed by "language-entry rules," internal inference rules, and "language-exit rules." But the construction of such a system of rules is not at all what people normally have in mind when they speak of a word's "conceptual role" or "use." (It was certainly not what Wittgenstein had in mind!) Indeed, everything I so far conceded as to the importance of "conceptual role" had to do with the importance of conceptual role in this intuitive sense; but in so doing, I conceded too much; for what has that sense to do with Block's technical notion?

I have remarked a number of times that to identify meaning with "conceptual role" would amount to a total change of topic, and not to an account of meaning. If this remark has seemed too harsh to some of my readers, it is, I think, because they too have in mind what I have just called "conceptual role in the intuitive sense." To identify meaning with conceptual role in the intuitive sense would not be as total a change as to identify it with the *syntactic or procedural* notion that Block has in mind; but it would also be a mistake. It would be a mistake because the identification of conceptual roles in the intuitive sense (that is, the identification of "central" beliefs, inferences, etc.) *presupposes* our ordinary ways of identifying beliefs as the same or different; that is, it presupposes the ordinary notion of meaning. It is not a substitute for that notion.

We can now see what has happened. Conceptual role semantics is offered as a *defense* of the claim that computational psychology is possible, that is, the claim that one can give a computational analysis of such relations as sameness (or "similarity") of meaning. The way in which this is to be done, however, is left so vague that the slogan "Conceptual Role Semantics" has in fact no content over and above the slogan of "functionalism" itself. Block is simply saying we shall *somehow* succeed in giving a functionalist account of *something* which when combined with a functionalist account of reference will yield a theory of meaning. The something is called "conceptual role," and is supposed to capture the fact that certain beliefs are central and the fact that certain inferences are central. But, as we have seen, there are a host of counterexamples to the claim that sameness and difference of meaning are a matter of conceptual role in *that* sense. Indeed, our whole discussion began with the fact that meanings remain invariant under enormous changes in conceptual role in *that* sense. Block's fallback position is to say, "Well, if it is only certain features of conceptual role, and not conceptual role as a whole, that fix meaning, that's a problem for any theory of meaning." But the problem—the problem "for any theory of meaning"—is just the problem for which *this* theory of meaning was offered as a solution!

Concluding Remarks

Even if Mentalese exists, we have seen that to identify the meaning of a word or sentence (or the "narrow" component of meaning) with "the corresponding formula in Mentalese" cannot be right; the meaning of a symbol cannot simply be another symbol, even a symbol in "brain writing." Nor can it be the "observable properties" (or perceptual prototype) associated with the symbol in Mentalese. And the suggestion that the meaning (or its "narrow" component) is the "conceptual role" of the formula in Mentalese associated with the symbol is wrong if conceptual role is taken in a Sellarsian sense, and useless otherwise. We are left without any serious candidate for the role of the mental object that traditional theories postulate.

What alternative do I suggest, then?

If to suggest an "alternative" is to *accept* the narrow content/broad content dichotomy and to offer an alternative *reductionist* account of what "narrow content" is—to suggest a story about how narrow content is definable in terms of the syntactical properties of "mental representations" (including the ones associated with "perceptual modules")—then I have no alternative to suggest. I come to bury the narrow content/broad content distinction, not to rescue it, and I come to bury the idea that an account of meaning must be reductionist. Not only is there no positive reason to believe that such a reductionist account exists to be found; the properties of the way we actually use the notion of meaning—the features (meaning holism, invariance under belief fixation) discussed in the first chapter, as well as the features (the centrality of reference, the contribution of the environment, and the division of linguistic labor) discussed in the present chapter—militate against the whole division of meaning into a "narrow content" and a "broad content." The motivation for that division was to rescue something of the traditional (Aristotelian) account, in particular to rescue the picture of meanings as isolable objects in people's heads. But that isn't—that can't be—the way language works.

Saying that I don't have an "alternative to suggest" in *those* terms is not at all to say that there is nothing to be said about meaning, or that the notion is "simple and unanalyzable." The reader—at least the sympathetic reader—will have noticed that I *have* all along been saying things about meaning. Moreover, those things are, one and all, things that the traditional picture kept us from noticing. That reference is not fixed by mental representations is conceded by the theories I have been criticizing (with the exception of Searle's), but it was not *discovered* by them. The social dimension of meaning—the division of linguistic labor—is still completely ignored by mentalistic theories. And the fact that stereotypes ("A king rules a country," "A game has winners and losers," "A witch has magical powers as a result of a pact with the Devil") play a much more important role than do *analytic* truths is likewise ignored. To be sure, the division of linguistic labor is a phenomenon described using the notion of *reference*, the notion of a stereotype belongs to the level of propositional-attitude psychology—none of the *informative* things we can say about meaning is at the computational or at the neuronal or at any other *reductionist* level. But so what?

We have all seen one social or human science after another—psychology, sociology, economics—come under the sway of some fad. In the United States such fads were more often than not the product of a reductionist idea of what it means to be "scientific." The idea that nothing counts as a contribution to "cognitive science" unless it is presented in terms of "mental representations" (and these are described "computationally") is just another case of this unfortunate tendency.

Chapter 4
Are There Such Things as Reference and Truth?

W. V. Quine would no doubt say that the difficulties with "mentalism" pointed out in the preceding chapters are grist for his mill. Quine has long contended[1] that it is a mistake to think of meanings as objects in the head; that the notion of synonymy is hopelessly vague, at least if our interests are theoretical; and that even reference is "inscrutable" when we are dealing with a language other than our own. In our own language, after it has been translated into a "regimented notation" (i.e., predicate calculus), it is easy to specify the extension of "refers to." The "disquotational" truths that "cat" refers to cats, "electron" refers to electrons, "superego" refers to superegos, etc., suffice to determine the extension of the relation "refers to" when that relation is restricted to the "home language." Every question of *sameness* of reference—e.g., "Does 'water' have the same reference as 'H₂O'?"—just reduces to the corresponding first-order question ("Is water H₂O?").

To the objection that translation into a regimented notation presupposes some notion of sameness of meaning, Quine's response is that such translation is a "free creation," not a discovery of some content already there. Quine thinks of unregimented natural language as a system of noises which has the function of helping us to anticipate "stimulations of our nerve endings," but to which no serious scientific notion of *reference* or *truth* is applicable.

To the objection that any theory of the propositional attitudes presupposes that we can compare utterances made in different languages (or thoughts "subvocalized" in different languages) for sameness or difference of meaning, Quine's response is that the propositional attitudes belong to folk psychology, not to science. We should settle for behaviorism, in Quine's view.

I cannot discuss here the deep philosophical reasons that Quine offers in support of these views. My plan in this book is to evaluate the prospects of a particular philosophical research program, or rather of a family of research programs, without allowing the discussion to broaden into a discussion of metaphysical issues in general.

Of course this is not completely possible; every philosophical program touches on deep metaphysical issues. But there is a fundamental difference in what may be called "level" between the question
"Can a reductionist account of broad content (respectively narrow
content) be given?" and such questions as "Is there a language-
independent world at all?" Here I intend to confine myself largely to
questions at what may be considered the "shallower" level; questions
that face philosophers who consider themselves "scientific realists."
What bearing the results we obtain at the shallower, or less metaphysical, level have on contemporary discussions in metaphysics will
be briefly discussed in the final chapter; but for the most part, this
question will be left for a future book.

Still, I did promise at the very beginning of this book that I would
say something about the "eliminationist" line toward the propositional attitudes; that is, about the view that it is not a disaster if a
reductionist account of the propositional attitudes turns out to be impossible, because the propositional attitudes (and especially belief
and desire) are quasi-mythological entities anyway, part of a body of
superstition called "folk psychology" (by Stephen Stich[2] and by Paul
and Patricia Churchland,[3] for example).

As I understand it, these "eliminationist" philosophers are not
committed to the "elimination" (from serious scientific or philosophical talk) of what is called *de dicto* belief; that is, to "eliminating" the
idea that people sometimes *affirm sentences*. Although the Churchlands, in particular, seem to think that only explanation in terms of
the functioning of neurons is "really" explanation, Stich, at least, is
friendly to computational accounts of the functioning of the mind/
brain. An account according to which "holding true" the sentence
"There are a lot of cats in the neighborhood" turns out to be, say,
"entering" a representation of that sentence in a "belief register"
would not be ruled out by Stich's version of eliminationism, at least.
What would be ruled out (more precisely: what Stich is extremely
skeptical about) is the idea that someone who believes "There are a
lot of cats in the neighborhood" and someone who believes "Yesh
harbe chatulim beshkuna" (Hebrew for "There are a lot of cats in the
neighborhood") are in *any* common physical or computational state
at all. In short, that there is a *state which is independent of one's local
natural language* which all humans who have the "propositional attitude" of "believing there are a lot of cats in the neighborhood" are
in, is, Stich maintains, almost certainly false.

In the chapters which follow, I shall offer reasons for believing that
Stich is, so far, right. But Stich concludes (and the Churchlands con-

clude, though for somewhat different reasons) that *there is no such thing* as a propositional attitude; there is no such thing as believing that there are a lot of cats in the neighborhood. And here I cannot follow these philosophers.

Why "Folk Psychology" and Not "Folk Logic"?

The central problem with the Stich-Churchland line is easily stated. Their whole argument turns on the following inference: if the instances of X do not have something in common which is *scientifically* describable (where the paradigm science is neurobiology in the case of the Churchlands and computer science in the case of Stich), then X is a "mythological" entity. There is, however, no attempt to apply this attitude *consistently*. For example: Suppose these philosophers are right, and there are no such things as *desires* or *purposes*. What makes various things all members of the class Chair is that they are portable *seats for one person* (with a back). Being a *seat for one person* is just being *manufactured for the purpose of being sat upon by one person at a time*. If there are no "purposes," then it is "mythology" that all chairs have something in common. So not only are there no such things as beliefs, if this view is right; there are no such things as chairs! Again, the Churchlands at least seem happy to talk of *causation* and to employ subjunctive conditionals; but there is no reason at all to think that all instances of "bringing about" or all instances of a particular "disposition" have anything in common which is scientifically describable.[4] So, in fact, the "eliminationism" of these philosophers is highly selective!

Nowhere is this selectivity more apparent than in the silence of both Stich and the Churchlands with respect to notions of *extensional* semantics—the notions of reference and truth. What is surprising about this silence is that the great pioneer of the eliminationist line— Quine himself—has long emphasized that the same difficulties beset the notions of intensional and of extensional semantics once we leave the confines of our "home language." If I "ascribe a propositional attitude" to little Amos in Tel Aviv, who believes *yesh harbe chatulim beshkuna*, by saying that he "believes that there are a lot of cats in the neighborhood," I am employing *translation;* that is, I am tacitly invoking the notion of *synonymy*. And if I say that Amos's belief is *true*, I am employing my own belief that there are a lot of cats in Ramat Aviv, where Amos lives, *and also employing my translation of the sentence by which Amos expresses his belief*. To classify the utterances of speakers of other languages as "true" and "false," we must not only have beliefs of our own about the world—those we need to do *anything*—we must also have a way of translating the utterances of those speakers into a language we ourselves understand. Similarly, if I say that Amos's word *chatulim* refers to cats, I am relying on the translation

    chatulim means *cats.*

If notions which depend on translation are to be rejected, then it seems we should reject the notions of *reference* and *truth* (unless these are to be restricted to words and sentences in English); the idea that there are properties of reference and truth (or falsity) possessed by words and sentences in anything that deserves to be called a language would appear to be as much of a myth as the idea that there are "propositional attitudes." But reference and truth are the fundamental notions of *the* fundamental exact science: the science of logic. Why don't the eliminationists speak of "folk logic" as well as of "folk psychology"?

I once put just this question to Paul Churchland,[5] and he replied, "I don't know what the successor concept [to the notion of truth— H. P.] will be." This is honest enough! Churchland is aware that the notion of *truth* is in as "bad shape" as the notion of *belief* from his point of view, and accepts the consequence: we must replace the "folk" notion of truth by a more scientific notion. But the innocent reader of Churchland's writings is hardly aware that he is also being asked to reject the classical notion of truth!

Disquotation, Anyone?

Anyone reading philosophical literature today—especially literature pro and con "scientific realism"—must be aware that many thinkers hold that the problems of reference and truth have been solved by something called the disquotational theory of truth. "The disquotational theory," mind you! Curiously enough, a *locus classicus* is rarely cited. As one reads further in this literature, one encounters all of the following claims: (1) The disquotational theory is antirealist. (2) The disquotational theory is an alternative—some would say *the* alternative—to the classical realist "correspondence theory of truth." (3) The disquotational theory is indeed the alternative to the correspondence theory, but *not* [as claimed by (1)] incompatible with realism. (4) The disquotational theory, far from being incompatible with the correspondence theory [as claimed by (2)], is a rational reconstruction of the correspondence theory.

The explanation of this state of what can only be called complete confusion is that the various authors who speak of "*the* disquotational theory of truth" do not all have the same theory in mind (and some of them probably have *no* definite theory in mind at all). On the one hand, many people who speak of a "disquotational theory of truth" have in mind Tarski's theory[6] (which Tarski himself called "the semantical conception of truth"), a theory which has also been defended by Carnap and by Quine. Tarski said, in various places, that this theory captures what is correct in the "correspondence theory" (hence view 4), *and* that the semantical conception of truth is *neutral* with respect to the realism/antirealism issue. On the other hand, other people seem to have in mind a theory that occurs in Ayer's *Language, Truth and Logic*—a theory that has been called the "redundancy theory" or the "disappearance theory" of truth (and one which Tarski himself explicitly *rejected!*). These disquotational theories of truth are relevant to the topic of eliminationism for the following reason: if reference and truth can indeed be explicated without any reference to psychology at all, then reference and truth are not a problem for "cognitive science." If they can be explicated in nonpsychological terms, then giving up folk psychology does *not* require giving up the notions of reference and truth. The eliminationist does not have to tamper with logic.

Let me begin by considering the "semantical conception of truth" as expounded by Carnap (following, of course, Tarski). In Carnap's account there was no thought of either "eliminating" the notion of truth or confining it to one's "home language"; Carnap was convinced that Tarski had *legitimized* the notion of truth (by showing us how to reduce it to notions which are themselves uncontroversial).

The "Semantical Conception" of Truth

I first met Rudolf Carnap in 1953. I had just come to Princeton as an assistant professor, and Carnap was spending two or three years at the Institute for Advanced Studies. In spite of the differences in our ages and "status" (I was an almost brand new Ph.D., and Carnap, of course, was a world famous philosopher)—neither of which meant anything at all to Carnap, who was totally indifferent to everything except the importance of philosophy, and very generous of his time— we became friends, and the hours I spent in the little house which the Institute had provided for Carnap, as well as for Ina (who called her husband simply "Carnap," as did all his friends) and for Marny, the giant and ferocious German shepherd, were as happy as any I can remember. One of my discussions with Carnap concerned the "semantical conception" of truth, and I regret that I was too timid to press my doubts to the end in that discussion.

In various books, Carnap had illustrated Tarski's theory by using the exceptionally transparent example of a language with finitely many sentences. (The use of very simple examples to make very difficult philosophical points was one of Carnap's great gifts.) Thus, if L_1 is a language with just two sentences, say,

> *Der Mond ist blau* (meaning The moon is blue)

and

> *Schnee ist weiss* (meaning Snow is white),

then, according to Carnap, one could define *S is true in L_1* thus:

> *S* is true in L_1 if and only if {(*S* = "Der Mond ist blau" and the moon is blue) or (*S* = "Schnee ist weiss" and snow is white)}.

In the same way, one could define *W refers to x in L_1* thus:

> *W* refers to *x* in L_1 if and only if {(*W* = "Der Mond" and *x* = the moon) or (*W* = "Schnee" and *x* = the substance snow) or (*W* = "blau" and *x* is blue) or (*W* = "weiss" and *x* is white)}.

When the language has infinitely many sentences, the simple technique of giving the truth definition in the form of a *list* will not work, but we know from the work of Tarski that such a definition can nevertheless be given. That is, it is possible to define "true in German," where German is, say, a suitably formalized version of the language Germans speak, in such a way that it is a *logical consequence* of the definition that

> "Schnee ist weiss" is true in German if and only if snow is white.

In the case of the simple example, we saw very clearly what is going on (this testifies to Carnap's expository brilliance): "true in L_1" has been identified with the property a sentence has if (case 1) it is spelled *D-e-r-space-M-o-n-d-space-i-s-t-space-b-l-a-u* and the moon is blue or (case 2) if it is spelled *S-c-h-n-e-e-space-i-s-t-space-w-e-i-s-s* and snow is white. So of course the sentence with the spelling *Schnee ist weiss* is true in L_1 if and only if snow is white—this is *logically necessary* given this definition of "true in L_1."

Tarski's way of defining "true in *L*," where *L* is whichever formalized language you please (say, English, if we have succeeded in formalizing English), is much more difficult to explain, because the languages Tarski considers have infinitely many sentences; but the result is the same. In effect, to be "true in L" is identified with the property of having the spelling of any one of the sentences of L (say the nth, in some standard enumeration) *and* its being the case that the nth condition in a list of truth conditions recursively associated with the sentences of L by the definition Tarski constructs is satisfied.

The objection I raised in my conversation with Carnap was this: it *isn't* a logical truth that the word "Schnee"—that is, the sequence of marks S-c-h-n-e-e—refers to the substance snow, nor is it a logical truth that the sentence "Schnee ist weiss"—that is, the sequence of marks S-c-h-n-e-e-space-i-s-t-space-w-e-i-s-s—is true in German if and only if snow is white. Obviously, if the history of the Indo-European languages had been just slightly different, "Schnee" might have ended up denoting water, and in that case "Schnee ist weiss" would not have been true in German—although it would be true in German according to *this* definition of "true in German."

Carnap's reply was that "everything depends on the way the name of the language—'German' or whatever—is defined." If by "German" we mean "the language spoken by the majority of the people in Germany" or "the language spoken by the people called 'Germans' in English," then it is only an *empirical* fact that "Schnee" refers to the substance snow in German, and only an *empirical* fact that "Schnee ist weiss" is true in German if and only if snow is white (i.e., my objection is correct). But in philosophy, Carnap urged, we should treat languages as abstract objects, and they should be identified (their names should be defined) *by their semantical rules.* When "German" is defined as "the language with such and such semantical rules," it *is* logically necessary that the truth condition for the sentence "Schnee ist weiss" in *German* is that snow is white.

I was not satisfied, but I did not continue the argument because I was too intimidated by the great presence. What I *thought* but did not say was: *And, pray, what semantical concepts will you use to state these "semantical rules"? And how will* those *concepts be defined?*

I can clarify my (unspoken) objection using Carnap's own mini-example. Suppose I want to define our little language L_1 (the one with just two sentences) "by its semantical rules." If I use the concept "true in L_1" to state those rules, the result will not be circular, as one might suppose from the presence of the name L_1 as a syntactic part of the predicate "true in L_1." It will not be circular because "L_1" does not actually occur as a part of the *definiens* of "true in L_1." But the result is interesting. First, however, let me write down the definition of L_1 using not the notion "true in L_1" but the intuitive notions of *reference* and *truth*—reference and truth in L, where L is a variable which can stand for any language at all:

Definition A

$L_1 =_{df}$ the language L such that, for every term W and for every x, W refers to x in L if and only if (case i) W = "Der Mond" and x = the moon, or (case ii) W = "Schnee" and x = the substance snow, or (case iii) W = "blau" and x is blue, or (case iv) W = "weiss" and x is white; and such that for any sentence S, S is true in L if and only if (case a) S is spelled D-e-r-space-M-o-n-d-space-i-s-t-space-b-l-a-u and the moon is blue or (case b) S is spelled S-c-h-n-e-e-space-i-s-t-space-w-e-i-s-s and snow is white; and (syntactic restriction) no inscription with any other spelling is a well-formed formula of L.

This is, indeed, a uniquely identifying description of the language we have called L_1—a definition in terms of L_1's semantical rules and syntactic properties. But it uses a *universal* notion of truth, a notion of truth in an arbitrary language ("truth in variable L"); and not only did Tarski not show us how to define *this* notion, it was his contention that it *cannot* be defined without running afoul of the paradox of the Liar and the other famous semantical paradoxes. The whole point of Tarski's (and of Carnap's) position in semantics is to avoid using or countenancing any such universal notion of truth.

Suppose, then, we try to rewrite the above definition in terms of the notions we do have, the notions *reference in L_1* and *true in L_1*:

Definition B

$L_1 =_{df}$ the language L such that, for any term W and for any x, W refers-to-x-in-L_1 if and only if . . . (as before); and such that for any sentence S, S is true-in-L_1 if and only if . . . (as before).

This is no longer open to the foregoing objection. I have hyphenated "refers-to-x-in-L_1" and "true-in-L_1" to emphasize that what we are to imagine is that these words have been replaced by their definitions (the ones Carnap gave). This isn't circular, as I explained, because "L_1" does not occur in those definitions. But now something really odd happens. The above definition is *not* a uniquely identifying description of L_1, as we want it to be; in fact, *every* language which contains just the two sentences that L_1 contains, but with arbitrary meanings, satisfies *this* definition!

Suppose, for example, L_2 is a language in which the two sentences have the following meanings:

Description of L_2

"Der Mond ist blau" is true if and only if *the sky is blue*

"Schnee ist weiss" is true if and only if *water is white*

What Carnap wanted was a definition of L_1 "by its semantical rules"—a definition which achieves what the impermissible Definition A would have achieved, if it had not run afoul of the ban on universal notions of truth. Such a definition must be satisfied by L_1 but, of course, not by L_2. But Definition B *is* satisfied by L_2. To see this, observe what it comes to when we use the definition Carnap gave of the predicate "true-in-L_1":

Definition B (unabbreviated)

$L_1 =_{df}$ the language L such that, for any term W and for any x, {(W = "Der Mond" and x = the moon) or (W = "Schnee" and x = the substance snow) or (W = "blau" and x is blue) or (W = "weiss" and x is white)} if and only if (case i) W = "Der Mond" and x = the moon, or (case ii) W = "Schnee" and x = the substance snow, or (case iii) W = "blau" and x is blue, or (case iv) W = "weiss" and x is white; and such that for any sentence S, {(S is spelled "Der Mond ist blau" and the moon is blue) or (S is spelled "Schnee ist weiss" and snow is white)} if and only if (case a) S is spelled "Der Mond ist blau" and the moon is blue, or (case b) S is spelled "Schnee ist weiss" and snow is white; and (syntactic restriction) no inscription with any other spelling is a well-formed formula of L.

Apart from the syntactic restriction, this is now an empty (tautological) condition. Every language which satisfies the syntactic restriction satisfies this!

In sum, if we try to rescue the claim that the "semantical conception of truth" correctly analyzes the notions of reference and truth in German (or whatever language) by insisting that "German" (or whatever language) must be "defined by its semantical rules," then in order to state *those semantical rules* we require *not* the notions of reference and truth provided by the semantical conception, but the "universal" notions employed in Definition A—that is, the very notions that the semantical conception wishes to replace!

An alternative reply to the one Carnap gave would be to renounce the claim that the semantical conception captures our semantic intuitions about truth and reference at all. Even if, as I have just shown, the semantical conception has the counterintuitive consequence that it turns out to be a logical truth that *"Schnee ist weiss" is true in German if and only if snow is white,*[7] still, definitions of this kind do give us *extensionally* correct notions of truth and reference; and we should be content with that much, one might argue.

But this is very different from claiming that our intuitive notion of truth has in any sense been analyzed. What is bizarre about these Tarskian "truth definitions" is that so many factors which are obviously relevant to the meaning of a sentence (and hence to whether the sentence is true or false) do not appear in the definition at all: under what circumstances it is considered correct to assert the sentence; what typically causes experts and/or ordinary speakers to utter the sentence; how the sentence came into the language; how a speaker typically acquires the use of these words; etc. If we accept the definitions Carnap gave of reference in L_1 and of truth in L_1, then whether a word refers to an entity in L_1 and whether or not a sentence is true in L_1, depend on how things are (whether or not the moon is blue and whether or not snow is white) and on how the sentence in question is *spelled*, but not on what the sentence *means*.

To see that this is so, observe that whether a sentence has the property "S is spelled S-c-h-n-e-e-space-i-s-t-space-w-e-i-s-s & snow is white" does not at all depend on what that sentence *means*. But to be "true in L_1" was defined as to have the disjunction of this property and another similar property. Occasionally a philosopher of a Tarskian bent seems to be dimly aware of this problem, and then the philosopher is likely to say, "Well, if you change the meaning of the words, then you are changing the language. Then of course you have to give a different truth definition." (Note that this is just what Carnap said, in a less formal guise.) But what is "the language"?

Donald Davidson has said that Tarski's theory makes truth relative to the language, because it replaces the intuitive "universal" notion of truth with an infinite series of notions, "true in L_1," "true in L_2," . . . But that is not the same thing as *representing the way in which truth depends on the language* (i.e., on the meanings of the words). For, appearances to the contrary, the term "L_{17}," or whatever, denoting the language in question, does not occur at all in the truth predicate. It is only the custom of *abbreviating* the truth predicate with an expression which does contain the name of the language that *gives the impression that the relativity of the notion of truth to the language is somehow captured in the truth definition*. But a look at the *unabbreviated* form of the definition reveals this to be an illusion.

Disquotation as Disappearance

Although no semantical notions are used in the Tarskian and Carnapian truth definitions themselves, a semantical notion *is* used in deciding when such a definition is correct, namely the notion of *translation*. When he constructed his little example, Carnap told us what the sentences of L_1 *meant;* that is to say, he *translated* them into natural language. And then he verified that the definition of "true in L_1" gave the desired results by proving the following two theorems (proving them in the natural language in which the definition was formulated; the natural language which functioned as the "metalanguage" in the case of his example):

> *Theorem 1.* "Schnee ist weiss" is true in L_1 if and only if snow is white.
> *Theorem 2.* "Der Mond ist blau" is true in L_1 if and only if the moon is blue.

Here Carnap was relying on a "Criterion of Adequacy" proposed by Tarski: a definition of "true-in-L," for a given formalized language L, is "adequate" if and only if all sentences of the following form are provable in the metalanguage in which the definition of "true-in-L" is constructed:

> "S" is true if and only if T (where S is any sentence of the object language and T is its translation into the metalanguage).

A little while ago, I said that some philosophers might urge us to accept Tarski's theory, not as an *analysis* of our intuitive notion of truth, but as a *replacement* for it. The need for the notion of "translation" in stating the Criterion of Adequacy is a problem for eliminationist philosophers who take this tack from Quine on. In the special case in which the object language is actually *contained* in the metalanguage, S can be taken to be its own translation, and the Criterion of Adequacy assumes the familiar form[8]:

> "S" is true in L if and only if S.

The fact that the sentence S is quoted on the left and used without quotes (or "disquoted") on the right is the "disquotational" character of the Criterion of Adequacy. The fact that the notion of translation is not needed in this special case—the case in which the object language is a part of the "home language," i.e., of the language which I use as my metalanguage—is the reason for Quine's finding the concepts of reference and truth unproblematic as applied to (suitable sublanguages of) his "home language."

   In the general case, there is no way of certifying that a truth definition is "adequate" without relying on the first version of the Criterion, and that version does employ the notion of translation. This has serious consequences for the claim that Tarski's account of truth is noncircular. For I argued in the previous chapters that the notion of meaning, and hence of translation, *presupposes* the notion of *reference*.

If this is right, then the Criterion of Adequacy employs a notion which belongs to the circle of notions Tarski wants to explicate.

For eliminationist philosophers of mind, "synonymy" is a notion to be rejected along with "folk psychology," and "translation" is simply a heuristic practice without any scientific (read: objective) status. Quine, at least, is willing to accept the consequences: truth and reference apply to languages other than English only when relativized to a particular translation scheme—and translation schemes are not *objectively* correct or incorrect! Quine's famous doctrine of the Indeterminacy of Translation implies that there is no objective sense in which the sentences of other languages can be thought of as possessing properties of truth and falsity apart from our fundamentally *subjective* ways of translating them into English (or whatever our home language happens to be).

In Tarski's theory, it is only the Criterion of Adequacy that has a disquotational character; the truth definitions themselves (the definitions of "true-in-L_1," "true-in-L_2," . . . for various languages L_i) are, as was said above, complex set-theoretic ways of saying that the nth sentence of the language (in a fixed enumeration) is true just in case the nth condition in a recursively generated list of conditions holds.

Philosophers who advocate what I shall call a "disappearance" theory of truth take the disquotational property of truth (the property illustrated by the fact that "Snow is white" is true in English if and only if snow is white) not as a Criterion of Adequacy to be satisfied by definitions of the property "true-in-L_1" for various languages L_i, but as itself the analysis of the notion of truth. But their purpose is the same as Tarski's: to provide an analysis of truth which is in terms so uncontroversial as to convince any philosopher (be he a physicalist, a phenomenalist, or whatever) that he should have no further qualms about using the notion (and, in fact, to show that the notion is, appearances to the contrary, totally "nonmetaphysical").

The disappearance theory is easily stated: *to say that a sentence S is true is not to ascribe a property to S at all, but simply to "affirm" S.* To say that "Snow is white" is true is not to ascribe a property to the sentence "Snow is white," but simply to affirm that snow is white. If we do not restrict this theory to formalized languages of the usual kind, then counterexamples are easy to come by: If S contains an indexical word (e.g., "I" or "now"), then, in general, when someone other than the person who said S says "S is true," he is *not* saying what he would be saying if he said S. If you say, "I am going to drive this car," and I say, "That's true," that is very different from my saying, "I am going to drive this car." (Usually this need to restrict the theory to regimented languages is not noticed, sad to say.)

The problem (which Paul Churchland was very aware of, in our conversations) is this: we think of language as a *rule-governed* system of practices. If I try to describe, say, the contents of this room, my statements (e.g., "There is an FM receiver in the room") may be successful or unsuccessful; and what counts as success is determined by conventions maintained by a community, not by my whim of the moment. To say of a description that it is "successful," in this normative sense, one normally uses some such word as "true," or "correct," or "right." But if "true" is not a property at all, then, in particular, it is not a *normative* property. Churchland has made it clear that he does think that there is a normative property which statements can have and fail to have;[9] his suggestion is that we were wrong in thinking that the classical notion of "truth" was the property in question.

A similar suggestion has been made by Richard Rorty in several well-known books. For Rorty, "is true" is just a "compliment" we pay to sentences we agree with. The substantive properties we want our sentences to have are (1) being "correct" by the standards of our cultural peers (this is the property we are interested in in "normal" discourse) and (2) enabling us to "cope." Rorty, however, is not an eliminationist. There is no reason why propositional-attitude talk cannot, for example, be "correct by the standards of my cultural peers," and it frequently helps us to "cope." Cultural relativism of the Rortian variety is antagonistic to the strain in "scientific realism" which holds that only science tells us what is "really" there; in fact, it is antagonistic to the very notion of being "really" there.

A suggestion which is more in the spirit of "scientific realism" is the following:[10] the realist who accepts the disappearance theory of truth need not, it has been suggested, give up the idea that words and sentences are *related* to things and happenings in the world. When one of my friends says to me, "I am going to buy a car," for example, I rely on the utterance as a sign that a certain event is likely to happen, much as I rely on a natural sign as an indicator that something is likely to happen. Perhaps what we want is that utterances should be *reliable indicators* of various happenings and states of affairs, rather than that they should be "true"? (Of course, if an utterance has whatever property we want successful utterances to have, then we will call it "true." This would explain our tendency to confuse whatever normative property we are aiming at, when we perform such speech acts as describing, with "truth"; and hence our tendency to think that truth must *be* a normative property. Only when philosophers start demanding a conceptual analysis of the notion of truth, producing counterexamples, etc., do we become aware that the normative property [or properties] we want utterances to have is [are] not "identical" with truth. And then truth starts to look "mysterious." Or so the disappearance theorist claims.)

The problem with this suggestion, as Hartry Field points out, is that there is no clear relation at all between the properties of sentence utterances as empirical indicators of and/or responses to events and the *formal structure of the language*. But an account of the normative properties of sentences which aims at explaining the contribution that linguistic practice makes to human social life must, one would think, be *systematic*; that is, it must exploit the syntactic and logical structures of those sentences in a recursive way. (To what Field says, I would only add that if such an account cannot be given, then the scientific realist who accepts the disappearance theory will be unable to explain why we should continue to use classical logic at all; he cannot say that we should use it because it "preserves truth," since his whole point is that "truth" is not the name of a property. Yet we do very often accept sentences simply because they *follow logically* from other sentences we accept.)

The problem facing the scientific realist in the philosophy of mind is simply this: if he gives up the idea that truth is a property, then he risks giving up his realism as well. This was, in fact, Rorty's trajectory; Rorty was at one time an eliminationist in the philosophy of mind, and has gone on to become an eliminationist with respect to truth and reference (or with respect to the idea that there are "interesting" notions of truth and reference)—as a result of which he has also gone on to eliminate the whole problematic of scientific realism! On the other hand, if the scientific realist defends his realism in the traditional way ("correspondence theory of truth," etc.), then he is open to the charge that he has not eliminated the intentional at all; for *reference* is a paradigmatic intentional notion. To reject the propositional attitudes on the ground that they cannot be reduced to physical/computational properties while keeping the relation of "reference" (which is in exactly the same shape, in this respect) is incoherent.

Two possibilities remain open. The scientific realist may abandon his eliminationism and try to develop an account of reference and/or the propositional attitudes in physical cum computational terms; this is the program of functionalism, and I shall devote the chapters which follow to an examination of this program. Or he may keep his eliminationism and try to meet the difficulties Field has raised. In this case, he cannot simply argue that the propositional attitudes belong to folk psychology and must therefore be given up; this purely negative stance will not suffice. He has now taken on a *positive* program, the program of developing what Churchland called a "successor concept to the notion of truth" and of showing that we can account for our linguistic and scientific practice (including the use of classical logic) in terms of this successor concept. I would only say that until this program is more than a gleam in the eyes of some scientific realists, I do not myself expect it to succeed any better than the mentalist program or the functionalist program. To me it seems that what we shall have to give up is the demand that all notions that we take seriously be reducible to the vocabulary and the conceptual apparatus of the exact sciences. I believe it is reductionism that is in trouble—not intentionality itself.

Chapter 5
Why Functionalism Didn't Work

Many years ago, I published a series of papers[1] in which I proposed a model of the mind which became widely known under the name "functionalism." According to this model, psychological states ("believing that p," "desiring that p," "considering whether p," etc.) are simply "computational states" of the brain. The proper way to think of the brain is as a digital computer. Our psychology is to be described as the software of this computer—its "functional organization."

According to the version of functionalism that I originally proposed, mental states can be defined in terms of Turing machine states and loadings of the memory (the paper tape of the Turing machine). I later rejected this account[2] on the ground that such a literal Turing machine-ism would not give a perspicuous representation of the psychology of human beings and animals. That argument was only an argument against one particular type of computational model, but the arguments of the preceding chapters constitute a more general reason why computational models of the brain/mind will not suffice for cognitive psychology. We cannot individuate concepts and beliefs without reference to the *environment*. Meanings aren't "in the head."

The upshot of our discussion for the philosophy of mind is that propositional attitudes, as philosophers call them—that is, such things as *believing that snow is white* and *feeling certain that the cat is on the mat*—are not "states" of the human brain and nervous system considered in isolation from the social and nonhuman environment. *A fortiori*, they are not "functional states"—that is, states definable in terms of parameters which would enter into a software description of the organism. *Functionalism, construed as the thesis that propositional attitudes are just computational states of the brain, cannot be correct.*

One way of trying to meet this objection was discussed in an earlier chapter. This is to say, "Yes, but a *component* of meaning, call it 'content,' and a similar *component* of the propositional attitudes *is* 'in the head,' and this component could be a computational state of the brain in the case of each 'meaning' and each 'propositional attitude.'" Certainly one must be in an appropriate physical state, an appropriate computational state, and so on, to believe that there is a cat on the mat, but not in the sense that there is *one* physical state or *one* computational state that one must be in to believe that there is a cat on the mat. It *was* an insight of functionalism to point out that the different physical states one might be in while believing that a cat is on the mat need not have anything "in common" that can be specified in physical/chemical terms. Just as this was an insight, so the upshot of our discussion here was that the different *computational* states one might be in while believing that a cat is on the mat need not have anything "in common" that can be specified in *computational* terms.

Speaking at the level of spontaneous phenomenology, it is undeniable that we perceive one another as "thinking that the weather is muggy," "believing that she will miss her train," and so on. These are phenomenologically *real* conditions. But as soon as we ask whether a Thai speaker who believes that a "meew" is on a mat is in the same "psychological state" as an English speaker who believes a "cat" is on a mat, we run out of spontaneous phenomenology and begin to babble our favorite "theory." The reason, I think, is that we look in the wrong place. Rather than thinking of the propositional attitudes as having a phenomenological reality which springs from the possibility of asking oneself if one really got the other person or the text right, one looks for a reduction of the propositional attitudes to something that counts as more "basic" in one's system of scientific metaphysics. One looks for something definable in nonintentional terms, something isolable by scientific procedures, something one can build a model of, something which will *explain* intentionality. And this—the "mental process"—is just what does not exist.

Sociofunctionalism

A way of trying to meet my arguments might be to extend the notion of a computational state by including aspects of the environment.[3] Why not think of the entire society of organisms together with an appropriate part of its physical environment as analogous to a computer, and seek to describe functional relations within this larger system? Why not seek to characterize reference, in particular, as a functional relation between representations used by organisms and things which may be either inside or outside those organisms? Perhaps one would have to use both computational notions *and* physical/chemical notions in the definition of reference; but the point is that one might, in some way, accept the chain of arguments which link meaning to reference and reference to entities (experts and paradigms) "outside the head" of the individual speaker without conceding that the intentional cannot be reduced to the nonintentional.

The arguments I just summarized were, it might be pointed out in this connection, arguments against locating meaning and reference inside the head. They are arguments against methodological solipsism. *But one can be a reductionist without being a methodological solipsist, after all.* Functionalism may have to become more complicated. We may have to speak of functional (and partly functional) properties of organisms-cum-environments and not just of functional properties of individual brains. But functionalism is not yet refuted.

This might be a plausible line to take if the only objection to functionalism stemmed from meaning's being partly fixed socially and partly fixed by the natures of external things. But there was another aspect to my argument, an aspect which was already present in my *Meaning and the Moral Sciences,* and which became central in the argument against sophisticated mentalism with which I began this work. *Meaning and reference depend on what I called "discounting differences in belief."* Functionalists like myself or David Lewis[4] recognize that an ascription of meaning to someone's "representations," an interpretation of someone's language (or thought-signs), must proceed simultaneously with the ascription of beliefs and desires to the person being interpreted. But the ascription can never, in practice, make the other's beliefs and desires come out the same as ours. We construe this word as meaning *plant,* that word as meaning *water,* this other word as meaning *gold,* in spite of the fact that the beliefs of the speakers we are interpreting, as discovered by this very interpretation (by the "translation manual," as Quine calls it), disagree with ours—perhaps disagree over the nature of plants, the nature of water, the nature of gold. When we ought to count two words as having the same meaning in spite of the difference between *their* beliefs and *our* beliefs that the very interpretation we are constructing requires us to posit, and when the beliefs we are attributing as the result of our translation are so bizarre as to require revision of the translation, is a question of "reasonableness." *A functionalist definition of synonymy and coreferentiality would formalize (and, probably, "rationally reconstruct") these intuitive judgments of reasonableness.* And this, I have argued, would be no easier to do than to survey human nature *in toto.* The idea of actually constructing such a definition of synonymy or coreferentiality is totally utopian.

To see how difficult it can be to perceive that two expressions are, in fact, not synonymous, it is instructive to reflect that it took trained analytic philosophers about fifty years to come up with convincing counterexamples to the claim that "X knows that *p*" is synonymous with "X has true justified belief that *p*." On the other hand, when we say that two words *are* synonymous—say, that "cat" in English and "meew" in Thai or "gorbeh" in Farsi are synonyms—we are making a *projection about projections;* for not only are we discounting the difference between "standard cats" and "Siamese cats" or "Persian cats," at least to the extent of classifying them all simply as "cats," but we are inferring from their verbal behavior that the Thai or the Iranian would count our "standard cats" as "meew" or "gorbeh" respectively (though hardly as "standard meew" or "standard gorbeh"). The difficulties are, in fact, so great (and controversial cases are so numerous) that some philosophers (notably Quine) have proposed to drop the notion of "meaning," at least from science, and to speak only of reference. But decisions on sameness of reference are still unavoidable, and these involve discounting differences of belief just as decisions on synonymy do (as Quine recognizes).

Few philosophers are afraid of being utopian, however. Suppose a functionalist were to say, "I agree that we are not able, and may never be able, to define 'reference' or 'coreferential' or 'synonymous' in functionalist terms—that is, we are not able to do it in practice. But it could be done *in principle,* and that is what is philosophically important." Here the argument that defining any of these terms is at least as hard as, let us say, constructing a successful symbolic inductive logic is no longer relevant. Questions of "as hard as" are irrelevant if what we wish to know is what is possible "in principle." What should we say here?

What "In Principle" Means Here

When I speak of "defining" reference and the various propositional attitudes, I am not, of course, thinking of finding an "analytic" definition, one which analyzes the "concept" or "meaning" of "refers to," "believes," "desires," and so on. No one any longer believes that semantic and propositional-attitude predicates are semantically or conceptually reducible to physicalistic (or computational) predicates. The question is whether these semantic and propositional-attitude properties and relations are "reducible" to physical-cum-computational properties and relations in the way in which (to use a familiar example) the temperature of an ideal gas is reducible to mean molecular kinetic energy.

When we say that temperature has been reduced to mean molecular kinetic energy, we are claiming more than just that "temperature" is coextensive with suitably measured mean molecular kinetic energy. These two magnitudes could be coextensive (equal in numerical value in all cases) even if temperature and mean molecular kinetic energy were distinct physical parameters (as long as the two parameters were one-to-one related by a suitable physical law). They could even be coextensive by accident (suppose temperature and mean molecular kinetic energy differed only in a class of physically possible cases which were not ever produced by investigators and which did not occur spontaneously in nature). When physicists say that temperature has been reduced to mean molecular kinetic energy, they are claiming that (1) the relation between mean molecular kinetic energy and temperature is *lawlike:* there is no physically possible situation in which a body has a certain temperature and does not have the corresponding mean molecular kinetic energy; (2) the laws which "unreduced" (or "phenomenological," in the physicists'—not the philosophers'—sense of "phenomenological") temperature was supposed to obey are approximately obeyed by mean molecular kinetic energy; and (3) the effects that "unreduced" temperature was supposed to explain are (to the extent that they really exist) explained by mean molecular kinetic energy. It is the second and third of these conditions that distinguish between correlation, even "lawlike" correlation (finding that one magnitude is a function of another), and reduction (finding that temperature *is* mean molecular kinetic energy).

Applied to the present case, these conditions tell us that we must not say, for example, that *reference* has been reduced to some physical/computational relation R (defined over organisms-cum-environments) unless (*a*) reference is coextensive with R in all physically possible systems—coextensive for all physically possible organisms and environments such that those organisms are capable of using language, referring, etc., in those environments; (*b*) R obeys (approximately) the "laws" that reference is supposed to obey in intuitive (or anthropological) belief about reference; and (*c*) the presence of R explains the effects (to the extent that they really exist) that the intuitive or anthropological notion of reference was supposed to explain. Merely finding a functional relation R which is coextensive with referring for those organisms that happen to refer (perhaps, by chance, there aren't any other than human beings) would not be enough.

If we require only that condition *a* be satisfied, then the problem of showing that such a relation R exists becomes trivial. For in each situation in which some organism refers, there is at least one physical and/or computational property which uniquely describes that situation. If the disjunction of physical/computational properties counts as a physical/computational property even when the disjunction is infinite, then one could (if one were blessed with omniscience) simply select all the situations in which physically possible organisms refer, select a physical and/or computational property uniquely characterizing each one of the situations one had selected, and then form the infinite disjunction of those properties, adding clauses specifying which part of the situation possessing the property is the referring expression and which part is the object or kind of object referred to, thus:

(*definition*) y refers to z = ($\exists x$) (x is an organism or group of organisms plus an environment & y,z are parts of x & the following disjunction holds: ($P1(x)$ and y is the token of "chat" which stands in the relation $R1$ to x and z is the class of cats) v ($P2(x)$ and y is the token of "Elektron" which stands in relation $R2$ to x and z is the class of electrons) v . . . v . . . v. . . .)

(one disjunct per situation, where Pi is a property selected to individuate the ith situation and Ri is a relation selected to distinguish the particular token we are interested in from any other tokens of the same type that may occur in that situation).

This would not be a true reduction of the relation of reference to physical/computational terms, but a mere *list* of all the cases in which a physically possible token refers to a physically possible object or class of objects. Such a "relation" as the one on the right side of the above "definition"—a relation which is given by an infinite list which is itself not constructed according to any rule given in finitely many words—cannot appear in the sorts of statements we call "laws," nor can it appear in "explanations." Requirements *b* and *c* specify that the "definiens" in an empirical reduction must be a property or relation which we can define in the vocabulary of the reducing discipline (allowing as parts of that vocabulary constants for appropriate mathematical objects, e.g., tensor or scalar constants and mathematical functions), where "define" has the normal sense of *define in finitely many words.*

Some philosophers believe that in addition to the notion of physical possibility there is another notion of possibility, metaphysical possibility, that we also possess. Thus, a world in which Newton's laws hold (and gravitational attraction travels instantaneously from anywhere to anywhere) violates the Principle of Special Relativity, and is hence physically impossible (by our present lights). But, these philosophers say, there really could have been a world in which Newton's laws held—and this is supposed to mean more than just that such a world (or rather, a *description* of such a world) violates no law of logic.

Such a world is "metaphysically possible," these philosophers say. And the "intension" of *reference* is not specified, according to these philosophers, until we specify the extension of "refers to" in all metaphysically possible worlds, not just in all physically possible ones.

If we take this point of view, then the claim that it is possible in principle to say what reference is in physical/computational terms should be taken to mean that one can define a physical/computational relation (in finitely many words, as just explained) whose extension in any metaphysically possible world coincides with that of "refers to." If we don't regard the notion of "metaphysical possibility" as sufficiently well founded to support such a demand, then we should only require that the extension of an acceptable *definiens* should coincide with that of "refers to" in any physically possible world. In the present discussion I shall assume this latter (weaker) requirement is the one imposed; if it cannot be met, then *a fortiori* neither can the stronger "metaphysical" requirement.

Although the requirements that an acceptable empirical statement of theoretical identity (e.g., "light *is* electromagnetic radiation of such-and-such wavelengths") has to meet are different from the ones imposed by the philosophers (the so-called "phenomenalists," such as C. I. Lewis and—at one time—Rudolf Carnap) who sought to show that material-thing language is translatable into sense-datum language, the considerations we have just reviewed are not wholly dissimilar to the issues that arose in the debate about phenomenalism. At first the phenomenalists were content to claim that material-thing sentences could be "translated" into *infinitely long* sense-datum sentences; however, it was very quickly pointed out that unless the translation were finite (or the infinitely long translation could be constructed according to a rule which was itself statable in finitely many words), then the issues whether the translation exists, whether it is correct, whether it is philosophically illuminating, and so on, would be essentially undiscussable. The antiphenomenalists said, in effect, "Put up or shut up."

In the same spirit, I am saying to the functionalists (including my former self), "Put up or shut up." However, the antiphenomenalists did not put all the burden of proof on the phenomenalists. Reichenbach, Carnap, Hempel, and Sellars gave principled reasons why a finite translation of material-thing language into sense-datum language was impossible. Even if these reasons fall short of a strict mathematical impossibility proof, they are enormously convincing, and this is the reason why there is not (as far as I know) a single phenomenalist left in the world today. In the same spirit, I am going to give principled reasons why a finite empirical definition of intentional relations and properties in terms of physical/computational relations and properties is impossible—reasons which fall short of a strict proof, but which are, I believe, nevertheless convincing.[5]

The Single-Computational-State Version of Functionalism

The simplest version of functionalism is one that has never actually been defended, as far as I know. This is the theory that each propositional attitude, each emotion, and so on, corresponds to a definite computational state. The identification is supposed to be species-independent: *believing that snow is white* is supposed to be the same computational state for all physically possible organisms capable of having that belief.

We have already seen that this version will not do. But let us examine it again, allowing this time that the computational states (or physical states, or partly physically specified and partly computationally specified states) may be states of Xs which are complexes of organisms and environments and not just single organisms. Although more sophisticated versions of functionalism actually exist (and will be treated below), we shall see that the arguments which defeat the one-state-per-propositional-attitude version contain the essential idea which will defeat the more sophisticated versions as well.

Consider the following model for a speaker-hearer of a natural language:[6] the "organism" is an information-processing system (it could be a robot) which possesses a way of assigning "degrees of confirmation" to the sentences in its "language of thought," and a "rational preference function" which (together with the degrees of confirmation) determines how it will act in any given situation. Certain semantic distinctions must be marked in any such model: for example, we can tell when a word is *acquired* by the fact that the "c-function" of the organism (the function which calculates the degrees of confirmation) and the rational preference function are *extended* to a new range of sentences. We can tell when a word is *ambiguous* by the fact that (in the underlying "language of thought") the word is marked by subscripts, or functionally equivalent devices, as, for example, "nap_1" (short sleep) and "nap_2" (nap of a rug). But if all we are given to go on is the current subjective probability metric (the current degrees of confirmation), the current desires (the current "utilities"), and the underlying c-function by which the current subjective probability metric was formed on the basis of experience, then at least the first two of these will be different even in the case of speakers whose meanings we are prepared to count as the same. In short (this was the point of a paper I published a few years ago),[7] there will be no discernible synonymy relation extractable from the model itself, nothing to mark the fact that when I say "bureaucrat" and you say "bureaucrat" we are uttering words with the same meaning.

The problem does not disappear even if we suppose (as Carnap did) that we should include information to the effect that certain sentences are marked *analytic* in the very description of a formal language. Even if certain sentences are marked *analytic* by the model, say "bureaucrats are officials in large institutions," unless I have a criterion of synonymy to tell me that when I say "official" and you say "official" we mean the same thing, and that when I say "institution" and you say "institution" we mean the same thing, I cannot conclude that "bureaucrat" has the same meaning for both of us from the fact that this *sentence* is analytic for both of us. If there were some stock of biologically innate and universal words—say, "observation terms"—such that all words were analytically definable in terms of these, then an analyticity definition and an identification of these linguistically universal basic concepts would solve our problem in the case of humans (we would still be no better off for an arbitrary "physically possible species," of course); but that way of solving the problem is ruled out by the fact of meaning holism, as we saw in chapter 2. (Moreover, the word may have the same meaning even if we have a *different* stock of "analytic" sentences. For example, someone who lives in a monarchy may have the sentence "People appointed to high positions by the king are officials" in his stock of "analytic" sentences, while someone who doesn't know what a king is but who is acquainted with presidents will have different "analytic" sentences about officials in his language, but this is not what we count as a difference in the meaning of "official.")

Finally, Quine's celebrated "gavagai" example shows that problems of synonymy can arise even at the level of observation terms. (Chinese speakers with whom I have talked—including sophisticated linguists—are not sure that Chinese distinguishes between a general name and the corresponding abstract singular term, e.g., between "rabbit" and "rabbithood." If their hesitation is well founded, then there may be no "fact of the matter" as to whether a certain Chinese character means "rabbit" or "rabbithood" or neither-of-the-foregoing.) In fact, sameness of "stimulus meaning" (Quine's substitute for the notion of sameness of "analytic meaning" in the case of observational vocabulary) is not even a necessary condition for synonymy, even in the case of observational terms. A Thai speaker may not associate the same stimulus meaning with "meew" that I do with "cat," but it is still reasonable to translate "meew" as "cat." ("Elm" in English and "Ulme" in German would still be synonyms even if "Ulme" were an observation term for Germans—they all learned to distinguish elms—and "elm" were not an observation term for English speakers.)

So far I have argued that in the sort of model of linguistic capacity that seems reasonable given the insights of Quine's meaning holism, there is no way to identify a computational state that is the same whenever any two people believe that there are a lot of cats in the neighborhood (or whatever). Even if the two people happen to speak the same language, they may have different stereotypes of a cat, different beliefs about the nature of cats, and so on (imagine two ancient Egyptians, one of whom believes cats are divine while the other does not). The problems that arise "in principle" become much worse if the two "people" may be members of different "physically possible species."

Even in the case of a single species, the "functional organization" may not be exactly the same for all members. The number of neurons in your brain is not exactly the same as the number of neurons in anyone else's brain, and neurologists tell us that no two brains are "wired" the same way. The "wiring" depends on the maturational history and environmental stimulation of the individual brain.

Still, many thinkers would suppose, with Noam Chomsky, that there is some "competence model" of the human brain to which all actual human brains can be regarded as approximating. This model would determine the "space" of possible computational states that can be ascribed to humans. The problem in the case of two different species is that in this case there is no reason to assume that the space of possible computational states is the same, or that either space can be "embedded" in the other.

Consider, for example, the crucial "belief fixation" component of the model (in the Carnap-Reichenbach model, this is the c-function or inductive logic). Even if we assume the species are ideally rational, in the technical sense of obeying the De-Finetti-Shimony-Carnap-Jeffrey axioms, this leaves an enormous amount of leeway for different inductive logics (as Carnap and Jeffrey point out). Carnap and Jeffrey introduce the concept of a "caution parameter"—a parameter which determines how rapidly or slowly the logic "learns from experience," as measured by how large a sample size the logic typically requires before it begins to give significant weight to an observed sample mean. Different inductive logics can assign different caution parameters. Different inductive logics can also assign different weights to analogy, and count different respects as respects of "similarity." In short, different inductive logics can impose different "prior probabilities." Granting that the need for survival potential will reduce the variability, we must remember that we are talking about all physically possible species in all physically possible environments—that is to say, about all the ways evolution (or whatever—some of these "species" will be artifacts, e.g., robots) might work to produce intelligent life, compatibly with physical law, not just about the way evolution actually happened to work in the one environment that actually exists.

For example, if the species is one whose members are very hard to damage, then they can afford to wait a long time before making an inductive generalization. Such a species might use an inductive logic with a very large "caution parameter." What properties it will be useful to count as "similarities" or respects of analogy will obviously depend upon the contingencies of the particular physical environment. Perhaps in a sufficiently peculiar physical environment a species that projected "funny" predicates (e.g., Nelson Goodman's famous predicate *grue*)[8] would do better than a species with our inductive prejudices. Computers that have to compute very different "analogies" or employ very different caution parameters (caution parameters which can themselves be different mathematical functions of the particular evidence e, not just different scalars) may have totally different descriptions, either in the Turing machine formalism or in any other formalism. The number of states may be different, the state transition rules may be different, and there is no reason why either machine should have a table which can be embedded in (or even mapped homomorphically into) the machine table of the other.

This point becomes clearer when we drop the (in any case false) assumption that actual organisms are "rational" in the decision-theoretic sense. It is well known that human beings are not rational in that sense,[9] and the "irrationalities" (which may, of course, be adaptive) of different physically possible sentient species will again depend on the particular species and on the particular environment to which that species is adapted. Yet these irrationalities, to the extent that they are species-specific and "wired into" the very structure of the computations performed by the "brains" of members of the species, would have to be represented in the computational description of any such species.

In sum, not only is it false that different humans are in one and the same computational state whenever they believe that there are a lot of cats in the neighborhood, or whatever, but members of different physically possible species who are sufficiently similar in their linguistic behavior in a range of environments to permit us to translate some of their utterances as meaning "there are a lot of cats in the neighborhood," or whatever, may have computational states that lie in an incomparable "space" of computational states. Even if their way of reasoning in some situations is "similar" to ours (when we view them with the aid of some translation manual that we succeed in constructing), this does not imply that the states or the algorithms are the same. The idea that there is *one* computational state that every being who believes that there are a lot of cats in the neighborhood (or whatever) *must* be in is false. (Or rather, it would be false if it had any meaning—remember that functionalists have abandoned the Turing machine formalism, and so far we have only the vaguest descriptions of what the computational formalism is supposed to be. Without a computational formalism, the notion of a "computational state" is without meaning.)

What about physical states? The reason for introducing functionalism in the first place was precisely the realization that we are not going to find any physical state (other than one defined by the sort of "infinite list" that we ruled out as "cheating") that all physically possible believers have to be in to have a given belief, or whatever. But now it emerges that the same thing is true of computational states. And (finite) conjunctions, disjunctions, etc., of physical and computational states will not help either. Physically possible sentient beings just come in too many "designs," physically and computationally speaking, for anything like "one computational state per propositional attitude" functionalism to be true.

Equivalence

I already said that I do not know of anyone who ever actually held the one-computational-state version of functionalism to be right. Let me now describe a version which I have seriously considered.

Fix, or rather imagine to be fixed, some definite formalism for computational theory—say, for definiteness, the Turing machine formalism. Although each Turing machine has its own "space" of machine states, still one can mathematically describe the totality of these machines *and* their associated "spaces" of machine states. One can define predicates which relate the states of different machines in different ways, and the notion of *computability* has been defined for such predicates. What is true in this respect of Turing machines is equally true of any other kind of machine that might be taken as a model in computational theory.

Now, suppose Ms. Jones is an English speaker, and suppose we wish to ascertain that Ms. Jones's word "cat" is synonymous with the Thai word "meew" (or with the word "meew" as used on a particular occasion by a particular Thai speaker). We have to know that the extension of the two terms is (at least vaguely) the same to even consider accepting the synonymy of the two terms, and this requires some knowledge of the actual nature of the beasties in Jones's environment that she (or experts upon whom she relies in doubtful cases) calls "cats" and some knowledge of the actual nature of the beasties that the Thai speaker (or experts upon whom she relies) calls "meew." Granted that this decision can involve enormously many factors— can involve not only the speech dispositions of Ms. Jones and her Thai counterpart, but also the speech dispositions of other members of the linguistic communities to which they belong, and information about the microstructure and evolutionary history of paradigm "cats" and paradigm "meew"—still, *if* we can make this decision and *we* are Turing machines, *then* the predicate "word W_1 as used in situation X_1 is synonymous with word W_2 as used in situation X_2" must be a predicate that a Turing machine can employ: a recursive predicate or at worst a "trial and error" predicate.[10]

This argument makes the basic empirical assumption on which functionalism depends, namely, that there is some class of computers (e.g., Turing machines or finite automata) in terms of which human beings can be "modeled." If we are willing to make this assumption, then the attractive feature of the argument is that it does not presuppose that the two situations being compared involve identical "machines." All that is necessary is that the entire situation—the speaker cum environment—be describable in some standardized language. In short, the problem we faced in the preceding section, that it makes no sense to speak of the "same computational state" when the speakers (or the speakers cum environments) are not machines of the same type, does not arise if what we are asking is, "Does a certain definable *equivalence relation R* (the relation of coreferentiality) hold between an element of the one situation and an element of the other?" States of different "machines" can lie in the same *equivalence class* under an arithmetical relation,[11] and so can situations defined in terms of such states. In short, moving from the requirement that the "states" of speakers with the same reference (or believers with the same belief) be identical to the requirement that they be *equivalent under some equivalence relation which is itself computable, or at least definable in the language of computational theory plus physical science*, gives us enormous additional leeway. What we have to see is whether this leeway will help.

Suppose (returning to the example of Ms. Jones and her Thai counterpart) that our biology assures us that the beasties that Ms. Jones takes to be paradigm "cats" are indeed various sorts of domestic felines (*Felis catus*), and that the same thing is true of the beasties her Thai counterpart takes to be paradigm "meew." This does not show that the extension of "cat" is the same as the extension of "meew," for several reasons. First—to be somewhat fanciful—it might be that Thai has an ontology of temporal slices rather than things. "Meew" might mean "cat slice." Second, even if we assume that English and Thai both cut the world up into "things," "animals," etc., the classification used by scientific biologists might not be one either Ms. Jones or her Thai counterpart employs. "Meew" might mean "Siamese cat," for example. We have to know a good deal about the Thai speaker's speech dispositions (or those of others to whom she defers linguistically) to know that she would count non-Siamese cats as "meew." What is at stake, as Quine and Davidson have emphasized (not to mention European hermeneuticists such as Gadamer), is the interpretation of the two discourses as wholes.

To interpret a language, one must, in general, have some idea of the theories and inference patterns which are common in the community which speaks that language. No one could determine what "spin" refers to in quantum mechanics, for example, without learning quantum mechanics, or what "negative charge" refers to without learning a certain amount of electrical theory, or what "inner product" refers to without learning a certain amount of mathematics. This creates two kinds of problem for the idea that coreferentiality or "synonymy" is theoretically identical with a computable (or at least computationally definable) relation over properly parametrized situations.

First of all, we know that terms in different theories can be coreferential. There are different theories about electrons, different theories about gravitational force, different theories about multiple sclerosis. Suppose I am interpreting a term T_1 in a theory A held by Martians and a term T_2 in a theory B held by Venusians. Suppose that the theories A and B are different, but not so different that the terms A and B could not conceivably refer to the same physical entity. Suppose further that the similarities between the two theories are sufficiently great that *if* the environments are such that the terms T_1, T_2 *are* coreferential, then a normal interpreter would not regard the difference in belief between the Martians and the Venusians as a difference in meaning.

In such a case, the decision on the synonymy of the terms T_1 and T_2 will turn on the decision on coreferentiality, and this will reduce to the decision on the two questions "What does T_1 refer to in the Martian environment?" and "What does T_2 refer to in the Venusian environment?" Answering these questions may require knowing (1) whether the two theories *A* and *B* are approximately true on Mars and Venus respectively, and (2) under what interpretations. If the theories are, say, cosmological theories, then determining whether either *A* or *B* is approximately true (as understood and used in its respective linguistic and scientific community) may require information about the rest of the universe. It might be that God Himself could not tell whether term T_1 refers to *anything* if He were allowed to use *only* information about Mars (or whether T_2 refers to anything if He were allowed to use only information about Venus). In short, the assumption that in principle one can tell what is being referred to by a term used in an environment from a sufficiently complete description of that environment in terms of some standardized set of physical and computational parameters is false *unless we widen the notion of the speaker's environment to include the entire physical universe.*

Second, any theory that "defines" coreferentiality and synonymy must, in some way, survey all possible theories. A theory that figures out what people (or physically possible extraterrestrials, robots, or whatever) are referring to when they speak of "spin," and that decides whether the notion of "spin" in terrestrial quantum mechanics is or is not the same notion as the notion of "grophth" in Sirian Mootrux mechanics, or an algorithm that would enable a Turing machine to make such a decision (or to reach it "in the limit") given a description of the "situations" on Earth and on Sirius, must, in some way, anticipate the processes of belief fixation on which the understanding of quantum mechanics (including the mathematics presupposed by quantum mechanics) and Mootrux mechanics (including the mathematics presupposed by Mootrux mechanics) depends. Certainly such an algorithm would have to do more than "simulate" an ability that human beings actually have. For no human being can follow all possible mathematics, all possible empirical science, etc. This point deserves further discussion, however.

Surveying Rationality

The fact that one cannot interpret a discourse unless one can follow it suggests that an algorithm which could interpret an arbitrary discourse would have to be "smart" enough to survey all the possible rational and semirational and not-too-far-from-rational-to-still-be-somehow-intelligible discourses that physically possible creatures could physically possibly construct. How likely is it that there is such an algorithm?

First of all, the restriction to physical possibility is not really helpful. As far as we know, physics does not rule out the possibility of an intelligent being that survives for n years for any finite n whatsoever. For example, some astronomers have suggested that a physically possible intelligent being might have a body that was a gas cloud of galactic size; the being would move with an incredible slowness, so that its time scale would be almost inconceivably slowed down by our standards, but such systems might have arbitrary complexity. The fact that such a being survives n years, for some large n, does not mean that it is "long-lived" by *its* (slowed-down) standards, of course; but it could also be incredibly long-lived by its standards. The point I mean this example to illustrate is that we do not *know* of any laws of physics which exclude any finite automaton whatsoever from being physically realized and from surviving for any finite number n of machine stages.

Let us begin by considering a somewhat less mind-boggling question. Can we hope to survey (and write down rules for interpreting, perhaps by "successive approximation") the reasoning and belief of all possible *human* beings and societies?

Let us recall that there is no one form in which all human beliefs are cast. The predicate calculus is often treated by philosophers as if *it* were the universal language; but to put beliefs expressed in a natural language into the predicate calculus format, one must first *interpret* them—that is, one must deal with the very problem we wish to solve. A theory of interpretation which works only after the beliefs to be interpreted have been translated into some "regimented notation" begs the question.

Moreover, the predicate calculus format itself has problems. What should the variables range over? Analytic philosophers have a preference for material objects and sense data; but there is no guarantee that every human language and sublanguage, including the specialized sublanguages of various professions (psychoanalysis, theology, sociology, cognitive science, mathematics . . .), will employ one of these standard ontologies. In fact, we know that the sublanguages just mentioned, at least, do not. Space-time points are another choice popular with philosophers; but to tell whether someone is quantifying over points in Newtonian space, or in space-time, or in Hilbert space, or in the space of Supergravitation theory . . . one again has to *interpret* his or her discourse. And it is not at all clear how to represent quantum mechanical discourse in the format of standard predicate calculus. I am not thinking of the possibility that quantum mechanics may best be understood in terms of a nonstandard logic (although that illustrates the point in a different way), but of the problem of interpreting quantum mechanics in its standard ("Copenhagen") presentation. Copenhagen theorists claim that quantum mechanics does not treat the world as consisting of objects and observer-independent properties, but rather treats it as consisting of two realms: a realm of "measuring apparatus," described by one ontology and one theory (classical physics), and a realm of "statistical states," described by vectors in Hilbert space and projection operators on Hilbert space. The "cut" between these two realms is not fixed, but is itself observer-dependent—something the predicate calculus format has no way of representing. Even if it turns out that quantum mechanics is being presented in the wrong way by its own practitioners, as many philosophers have thought (though without coming up with an agreed-upon better way), to interpret a discourse in existing quantum mechanics one must first realize that the language of those practitioners is of this "nonclassical" kind. What other languages of a "nonclassical" kind that science (or history, or literary criticism, or . . .) might use are waiting to be invented?

Experience tells us that no human society is unsurpassable. For any human society, there is a possible other society which is more sophisticated, which has modes of conceptualizing and describing things which members of the first society cannot understand without years of specialized study. What is often said is true, that all human languages are intertranslatable; but that does not mean that one can translate a current book in philosophy or a paper in clinical psychology or a lecture on quantum mechanics into the language of a primitive tribe without first coining a host of new technical terms in that language. It does not mean that we could tell any "smart" native what the book in philosophy or the paper in clinical psychology or the lecture on quantum mechanics "says" and have him understand (without years of study). Often enough we cannot even tell members of *our* linguistic community what these discourses "say" so that they will understand them well enough to explain them to others.

It would seem, then, that if there is a *theory* of all human discourse (and what else could a *definition* of synonymy be based upon?), only a god—or, at any rate, a being so much smarter than all human beings in all possible human societies that he could survey the totality of possible human modes of reasoning and conceptualization, as we can survey the "modes of behavioral arousal and sensitization" in a lower organism—could possibly write it down. To ask a human being in a time-bound human culture to survey all modes of human linguistic existence—including those that will transcend his own—is to ask for an impossible Archimedean point.

Chapter 6
Other Forms of Functionalism

To the difficulties with functionalism pointed out in the preceding chapter the obvious response is to say, "Very well, the difficulties you raise do show that we cannot hope to have the Master Algorithm for Interpretation in practice. But remember, we set out to discuss what is possible in principle. In principle there could be such an algorithm or such a theory even if no human being could understand it *in toto*."

One problem with this response is that even if one grants that there are some things which are true or false whether or not human beings could ever know that they are true or false—say, that there was a rabbit in the area where my house now stands at 12 noon on September 14, A.D. 1000 (Gregorian), or that there are intelligent beings in some other galaxy, or that the number of stars in star cluster A is the same as the number of stars in cluster B—still, we at least know what *sort* of thing would make these statements true or false. But if we ask whether there is an *ideal* theory of interpretation, or, at least, a class of adequate "rational reconstructions" of the procedure of interpretation, then it is not clear that we know what would make the answer "yes there is" or "no there isn't" true—not even if we say that the procedure is restricted to human beings (but not to any particular culture or time period). What is it to be "ideal" or "adequate," after all? Normally we think of the "adequacy" of a rational reconstruction as consisting in its agreement with some body of practice and some set of intuitions. *Whose* practice and whose intuitions is the Ideal Theory of Interpretation (or *an* Ideal Theory of Interpretation, if there is more than one candidate) supposed to agree with? Not those of human beings, since we have seen that (almost certainly) no human being could understand an ideal theory of interpretation that "worked" for arbitrary human cultures and languages. The practice and the intuitions of a rational being smarter than we? What would define "rational" for such a being? No doubt we could be sure that some being we might meet *is* smarter than a human in certain respects (as a child can be sure that a grownup is—in certain respects—

smarter than the child); but could we know that such a being *never* makes a mistake, even when he talks about matters as delicate as human intentions and human cultures? Are we just to accept the notion of such a being (or, perhaps, the notion of an "adequate rational reconstruction") as an *unreduced* notion? But then, why should we not just accept the propositional attitudes themselves as unreduced notions?

Another problem with this response is that the program of functionalism was to make sense of the propositional attitudes as *possible psychological states of any physically possible organism*. Functionalists criticized older forms of materialism on precisely the ground that they were "species-chauvinist." But if the program is to construct a theory which explains propositional attitudes, semantic notions, etc., over all possible species, then the problem we face is that our theory must "survey" the possible modes of conceptualization of all physically *possible* rational beings. Then the difficulty we just faced in connection with human beings arises in a worse form: if no human being could survey all the possible modes of conceptualization of other human beings in all the possible human languages and human cultures, then certainly no rational species could survey all the possible modes of *nonhuman* culture, language, or mode of conceptualization, including those of beings almost infinitely smarter than themselves. If there is a theory which states precise criteria for "correct interpretation" (whether in the sense of reference-preserving translation, sense-preserving translation, or correct paraphrase) for all possible rational species, a functionalist theory which is truly not "hydrogen-carbon-chauvinist," then *that* theory is one that no possible rational being (in the sense of "physically possible finite intelligent being") could understand. What the "correctness" of such a theory could consist in is as mysterious as the propositional attitudes in their most free-floating and unselfconscious applications.

The upshot of this discussion is that if computational psychology is to succeed in "reducing" the propositional attitudes—that is, if we are to find correct statements of theoretical identity in computational/physical terms in the case of the propositional attitudes—then mimicking (or "rationally reconstructing") actual procedures of interpretation is not the way to go about it. The idea of looking for a computable (or even a well-defined) equivalence relation between functional states which corresponds to the equivalence relation that the practice of "good interpreters" implicitly defines runs up against an insuperable difficulty in the inexhaustible open-endedness of the totality of conceptual schemes that have ("in principle") to be interpreted. We must look at other forms of functionalism.

Before doing this, there is one further remark that I wish to make. If one were willing to settle for a theory of interpretation which was species-specific and culture-specific, that is, for a theory which described precisely how correct interpretation should proceed for one particular language in one particular setting, then, if we made the "setting" sufficiently inflexible, such a theory might exist "in principle." But the disjunction of *all* such theories—or rather, the complex logical function of such theories which defines correct interpretation by saying (in effect), "If you are in culture 1 use theory A, if you are in culture 2 use theory B, if . . ."—would amount to just the sort of "infinite list" that we ruled out as not constituting a genuine reduction. Yet the considerations we have just surveyed—in particular, the open-endedness of what constitutes the "environment" of a species, the open-endedness of what constitutes a "form of discourse," and the fact that a theory which surveyed all "forms of discourse" (and the forms of belief fixation that correspond to those forms of discourse) would have to be able to "understand" the discourse of beings "smarter" that the beings who constructed the theory in question—suggest that there is absolutely no reason to think that a universal theory of interpretation would be anything but such an infinite logical function of species-and-culture-specific theories. The claim that there is something finitely specifiable that all cases of correct interpretation have in common is one that we have simply no reason to believe. And if propositional attitude assignment depends upon interpretation (as it is made to do in the proposal that we seek an equivalence relation between computational/physical states of organisms cum environments—an equivalence relation which formalizes "correct interpretation," and which counts token beliefs occurring in different organism-cum-environment complexes as "equivalent" just in case correct interpretative practice would dictate that one should interpret the two token beliefs as coming to the same thing), then the fact that the equivalence relation is defined by an "infinite list" (one not itself constructed according to any effective rule) means that the "state" corresponding to any propositional attitude (or, rather, the equivalence class of states) will likewise be defined by an infinite list which is not itself constructed by any effective rule. In short, the claim that "in principle a propositional attitude is an equivalence class of computational/physical states" is in as bad shape as the claim that "in principle talk of material objects is highly derived talk about sense data." The advocate of neither kind of reductionist claim is in a position to "put up."

David Lewis and I

Since the search for a computationally definable "equivalence relation" holding between computational states that "correspond" to the
same propositional attitude runs into these difficulties, the only re-
alistic option remaining open to a functionalist is to state the functionalist thesis in a way that does not depend on the computational
formalism—on the Turing machine formalism, or any of its successor
formalisms. Computational properties just don't seem to be what
"intentional systems" with the same propositional attitude have in
common.

One way of doing this has been suggested by David Lewis. In a
series of important papers,[1] Lewis has suggested that propositional
attitudes, experiences, and "mental states" are "implicitly defined by
a theory." By what theory? By a theory we already have, Lewis says:
the "platitudes" of folk psychology *already* constitute an implicit definition of all the "mental states" that we now speak about. Instead of
thinking about what a God's-eye theory of all physically possible intelligent organisms and automata might look like in computational/
physical terms, all we need to do is think about the folk psychology
we already possess.

> Think of common-sense psychology as a term-introducing scien-
> tific theory, though one invented long before there was any such
> institution as professional science. Collect all the platitudes you
> can think of regarding the causal relations of mental states, sen-
> sory stimuli, and motor responses. Perhaps we can think of
> them as having the form:
>> When someone is in so-and-so combination of mental states
>> and receives sensory stimuli of so-and-so kind, he tends
>> with so-and-so probability to be caused thereby to go into
>> so-and-so mental states and produce so-and-so motor
>> responses.
> Add also all the platitudes to the effect that one mental state
> falls under another—"toothache is a kind of pain," and the like.
> Perhaps there are platitudes of other forms as well. Include only
> platitudes which are common knowledge among us—everyone
> knows them, everyone knows that everyone knows them, and
> so on. For the meanings of our words are common knowledge,
> and I am going to claim that names of mental states derive their
> meanings from these platitudes.[2]

Lewis spells out his contention further with the aid of two notions
to which he helps himself: the notion of a "neurochemical state"[3] and

the notion of a "causal role." His idea is that each mental state is identical with a neurochemical state, but not with the *same* state in the case of different species. *Believing that snow is white* could be one "neural state"[4] in the case of Martians, and a different "neural state" in the case of human beings. Thus the search for a *species-independent* state to which a given propositional attitude (or a given "experience") can be reduced is given up. The pains, beliefs, etc., of Martians are not literally the same properties as the pains, beliefs, etc., of humans—not even when we would describe them with exactly the same words—but they have the same causal roles. What fixes the reference of propositional-attitude descriptions, sensation names, etc.? The fact that, as commonly understood, these sorts of descriptions and names designate states with certain causal roles, Lewis answers. If there are no such (neurochemical, or more broadly "physical") states, then we are *mistaken* in thinking we have experiences, beliefs, and so on.[5]

> From the postulate [the "conjunction of all these platitudes"] form the definition of the T-terms [the mental state terms]; it defines the mental states by reference to their causal relations to stimuli, responses, and each other. When we learn what sort of states occupy those causal roles definitive of the mental states, we will learn what the mental states are.[6]

If it is possible to find an ordered set of "physical states" which play the "causal roles" specified by a theory, when the T-terms ("the theoretical terms") of the theory are taken to designate the states in the set in the appropriate order, then that set of states is said to be a "realization" of the theory. In this terminology—which Lewis employs in these papers—Lewis is saying that any realization of folk psychology is an intended interpretation of folk psychology. The T-terms of folk psychology—the mental state terms—have more than one intended interpretation if there should happen to be more than one realization of folk psychology, and they have null denotation in the actual world if there are *no* realizations.

The problems with Lewis's account have to do with the critical notions of "physical state" and "causal role." In physics an arbitrary disjunction (finite or infinite) of so-called "maximal" states counts as a "physical state," where the maximal states (in classical physics) are complete specifications of the values of all the field variables at all the space-time points. If we count every sequence of physical states (in the sense just described) as a "realization" of every theory which comes out true when the T-terms are taken to designate the terms in the sequence (in some appropriate order), then every psychological

theory which has the sort of probabilistic-automaton structure that
Lewis's remark about the form of the "platitudes" suggests, and
which *correctly predicts the behavior of an object*, has a realization. The
requirement that a theory have a realization is too weak a requirement to serve Lewis's purposes, if "realizations" are allowed to involve arbitrary physical states. (For a proof, see the Appendix.) If this
notion of physical state were the one he intended, Lewis's theory
would come to the claim, which he explicitly rejects, that to "realize"
a psychological theory is just to *behave so that its predictions about
behavior come out true.* The difference between functionalism and behaviorism as positions in the philosophy of mind would entirely
disappear.

The result about realizations just alluded to (the result proved in
the Appendix) depends, however, on assuming that the "causal" relations that the program requires to obtain between the states it postulates are of the type that commonly obtains in mathematical
physics. That is, if a program says, as it might be, that "state A is
always followed by state B," then, in the proof in the Appendix, I
simply take this to mean that the function which determines the se-
quential relations of the states of the physical system in time given
the boundary conditions—in classical physics this would be determined from either the Hamiltonian or the Lagrangian equations of
the system—is such that any maximal state of the system which lies
in the region of phase space corresponding to state A and which is
compatible with the given boundary conditions and with physical
law will be followed by a maximal state which lies in the region of
phase space corresponding to state B. In familiar language, this is to
say that a mathematically omniscient being of the kind once envis-
aged by Laplace (the being who could deduce the whole future and
past of the universe given its maximal state at one instant of time and
the laws of physics) could *predict* that the system X would go into
state B at the relevant time given the information that it was in state
A at the earlier time and given the boundary conditions.

This is not the concept of causality that David Lewis has in mind,
however. In his papers[7] Lewis has proposed a different notion of
causality—one based on the notion of "similarity of possible worlds"
and on a theory of "events" which in turn presupposes his "theory
of universals."[8]

The main requirements in Lewis's theory of causation are (1) that
if we say "A caused B," we must be prepared to assert "If A had not
been the case, then B would not have occurred"; and (2) that A and
B must be the sort of predicates that pick out "events," where this
last requirement is not merely semantical but *metaphysical:* only pred-

icates which have a property Lewis calls "naturalness" or "eliteness"[9] pick out "events," according to Lewis.

In certain respects the notion of causal connection used in mathematical physics is less reasonable than the commonsense notion Lewis is trying to explain (or to provide with a metaphysical foundation). If, for example, under the given boundary conditions, a system has two possible trajectories—one in which Smith drops a stone on a glass *and* his face twitches at the same moment, and one in which he does not drop the stone and his face does not twitch—then "Mathematically Omniscient Jones" can predict, from just the boundary conditions and the law of the system, that *if* Smith (the glass breaker) twitches at time t_0, *then* the glass breaks at time t_1; and this relation is not distinguished, in the formalism that physicists use to represent dynamic processes, from the relation between Smith's dropping the stone at t_0 and the glass breaking at t_1. Lewis would say that there are possible worlds (with different boundary conditions or different initial conditions from the ones which obtained in the actual world) in which Smith does not twitch but does release the stone and the glass does break, and that these worlds are *more* similar to the actual world than those in which Smith does not twitch and also does not release the stone (assuming Smith *did* release the stone in the actual world).

Without going deeply into the mysteries of the possible-worlds explanation of counterfactual conditionals, one can sum this up as follows: when we consider what *would* have been the case if Smith had not twitched, we *keep* such things fixed as that he released the stone. This means that, in our ordinary use of counterfactuals, we consider (in some intuitive fashion) situations in which the *boundary conditions themselves* (or the initial conditions, or both) are quite other than they actually are. That is why a criterion of causality which considers ordinary-language counterfactuals can lead to quite different results than the mathematical physicist's criterion (and, in fact, the mathematical physicist's notion of causal relation is a quite unusual one from the standpoint of ordinary language as employed in nonspecialized contexts). Lewis's theory of similarities between possible worlds is an attempt to reconstruct the ordinary procedure of selecting appropriate hypothetical situations to consider in deciding on the truth or falsity of a counterfactual.

Similar remarks apply to the notion of an "event." Thus, let B be a mathematically well-behaved set of maximal states of a system S. Then "The state of S lies in B at time t" is a perfectly good description of an "event," from a physicist's point of view. In ordinary language, however, if the state-functions in the set B do not correspond to a "natural class," then we would not consider "The state of S lies in B at t" to be a genuine event type (or, at any rate, David Lewis would not consider this to be a genuine event type, and hence he would rule that "The state of S lies in B" is not a possible *cause* or *effect*). On the other hand, "An apple fell from a tree" *is* a possible cause or effect, *even though* the set of physical states in which an apple falls from a tree is highly disjunctive and badly behaved in a mathematical sense (it is not connected, not convex, may not be a Borel set or even a projective set, etc.).

After this digression, we can return to Lewis's theory: if what singles out the referents of the T-terms in folk psychology—the referents of the expressions which designate propositional attitudes, experiences, desires, and so on—is that these referents are "events" which also satisfy certain *counterfactual conditionals*, and all this is explained in terms of a primitive notion of "natural class" conjoined with a *similarity metric over possible worlds*, then Lewis's account is *not* a reduction of the propositional attitudes to anything physical, but rather a reduction of the propositional attitudes to a set of highly metaphysical properties and relations. The notions of "possible world," "similarity of possible worlds," and "naturalness" are surely in much worse shape than the notions of belief and desire that we are trying to explicate. (This objection applies, of course, to Lewis's explanation of the functioning of theoretical terms in general, and not just in psychology.)

Instead of using a notion of causation based on possible worlds and a metaphysical theory of "natural classes," one might stick to the mathematical physicist's notion, but refuse to allow disjunctions of maximal states which have "nothing in common" to be "realizations of the T-terms" of our theories. Apart from the question of a criterion for deciding which states have "nothing in common," we would then run into the problem of the previous chapter; for the whole burden of that chapter was that there is no reason to think that there *are* "states" (other than infinite disjunctions), in the sense of "neural" states, or "physical" states, or even "computational" states, which could be "realizations" of the propositional-attitude names and descriptions in folk psychology.

In "Philosophy and Our Mental Life"[10]—the last in my series of "functionalist" papers—I myself made a proposal which runs up against this same problem. Abandoning the idea that the Turing machine formalism would be a suitable model for the mind, I employed instead the (admittedly imprecise) notion of a "psychological theory." I defined two systems to be "functionally isomorphic" if there is a mapping of the "states" of the one onto the "states" of the other which makes them *isomorphic* models of that psychological theory. The "thesis" of functionalism became that all mental states (propositional attitudes, experiences, etc.) are preserved under functional isomorphism.

It was built into this proposal that something is a "model" of a psychological theory only if it has *nonpsychological*—physical or computational, or whatever—states which are related as the psychological theory *says* the mental states are related. The purpose of the notion of functional isomorphism was to avoid having to postulate that the physical state that fills the role of a given mental state must be the same in the case of different species or even in the case of different organisms. But it was still assumed that one can find one physical state per propositional attitude in the case of a single organism. Different organisms have only to be "functionally isomorphic"; they do not have to "realize" the appropriate psychological theory in the same way. In fact, there can be organisms which possess propositional attitudes and which are not even functionally isomorphic, according to the position I took in the paper in question, since I allowed that one can be an intentional system by being a model of *any* (appropriate) "psychological theory." (On the question whether models of *different* psychological theories could have the *same* mental state—say, the same belief or desire—this paper was silent.)

There are certain differences between the account in "Philosophy and Our Mental Life" and Lewis's account. On Lewis's account, a given propositional-attitude description, say, "believing that snow is white," refers to *different* physical properties in the case of different organisms. On my account, the propositional-attitude description was supposed to refer to the whole equivalence class of physical properties, rather than to one of the members of the class in the case of Oscar, a possibly different member of the class in the case of Elmer, etc. Again, on Lewis's account there is just *one* psychological theory we have to worry about, namely folk psychology (or common-sense psychology, as Lewis also calls it). The models of this one theory (properly reconstructed) are all the intentional systems there are, in Lewis's view. In my former view, we needed an ideal psychological theory, of a kind we do not presently possess, to really define what even one type of intentional system is. Nonetheless, the similarities between the two accounts are very striking (I am sorry to say that I did not see them clearly at the time). In both accounts the notion of being a model for a certain kind of theory is used to explain the notion of being a possessor of mental states. In both accounts the psychological theory simply replaces the computational formalism (Turing machines or whatever), as far as the job of "implicitly defining" the relevant mental states is concerned. And in both accounts the notion of being a "model" of a psychological theory is understood in the same way, as involving the possession of *physical* states which play the "roles" the mental states are alleged to play by the psychological theory.

If *any* physical state is allowed as a possible "realizer" for any "T-term" in a psychological theory, then the objection I made to Lewis also works against me: psychological theories will just have *too many* realizations if they have any (they will have none if they imply a false prediction). We must restrict the class of allowable realizers to disjunctions of basic physical states (disjunctions either of maximal states or of what I call "basic open" states in the Appendix) which really do (in an intuitive sense) have "something in common." Moreover, this "something in common" must itself be describable at a physical, or at worst a computational, level: if the disjuncts in a disjunction of maximal physical states have nothing in common that can be seen at the physical level and nothing in common that can be seen at the computational level, then to say they "have in common that they are all realizations of the propositional attitude A," where A is the very propositional attitude that we wish to *reduce,* would just be to cheat. So my former account needed it to be true that there is one physical or computational state per propositional attitude in the case of each single organism, if not in the case of each whole species, where the notion of a "physical or computational state" is itself understood in a "natural" way. But there is just no reason to think that there *is* one physical or computational state per propositional attitude, as we have already seen.

Not even when we consider just one single organism? No more than in the case of the whole human species. For consider: I myself started life as a speaker of French (my parents took me to France when I was eight months old). Later, when I came to the United States at the age of eight, I forgot my French and learned English (and a whole new set of beliefs, stereotypes, etc., in the process). There are people who change languages (and cultures) more than once in a lifetime, and who, as I largely did, forget their previous language and culture each time. There are people who have started life in a primitive tribal culture and who have ended up as Ph.D.'s in a sophisticated scientific culture. There can be as much difference in the belief sets, stereotypes, subjective probability metrics, etc., of one person at two different times as in the belief sets, stereotypes, etc., of two people in any two different human cultures. To survey all the states just one particular human being *could* be in while believing that *there are a lot of cats around here* is no less unbounded a task than to survey all human cultures and modes of belief fixation.

Lewis's Theory Further Examined

What we have just seen to hold in the case of my (former) account also holds for Lewis's account. His account too says that to have psychological states is just to be a "model" of a certain theory. He too needs to restrict the notion of a "realizer" so that highly disjunctive physical states (like the ones used in the proof in the Appendix) do not count as realizers even if they satisfy the postulates of folk psychology. Lewis does this by requiring the realizers to have the metaphysical property he calls naturalness or eliteness. But this means that his theory requires that each organism to which folk psychology applies be a model for folk psychology in just the sense that my account in "Philosophy and Our Mental Life" required—that is, that such an organism possess one physical state per propositional attitude. But we have seen that we are *not* models of propositional-attitude theory in *that* (highly reductionist) sense.

I can exploit my arguments to make a further point: folk psychology cannot play the explanatory role Lewis wishes it to play, the role of *defining* the propositional attitudes. If it could play this role, the "platitudes which are common knowledge among us" would already constitute a survey of all possible ways of having any belief that we can describe. And Lewis has not given us any reason to believe that they do this.

Consider the belief that (apart from dirt) snow is normally white. Suppose someone who lives in a future super-scientific culture (and who speaks a language unrelated to any present-day language) has a belief that should be so interpreted. (A problem in the radical interpretation of a language spoken by beings more sophisticated than we.) How would this be determined using beliefs of *our* culture such that "everyone knows them, everyone knows that everyone knows them, and so on"?

These future speakers might believe that snow is white on the basis of a quantum mechanical calculation. (Perhaps it snows so rarely that that is now the "normal" way of fixing the belief that snow is white.) So *our* platitudes about what "sensory stimulations" cause one to go into the mental state of believing that snow is white "with so-and-so probability" will not apply to these speakers. Imagine that they not only come to the belief in a different way than we, but that the sensory stimulations that suffice for us do not suffice for most of them, at least without lengthy calculation—they do not go immediately from the sensory stimulations to the "mental state."

To see how unlikely it is that we could break into the alien hermeneutical circle *merely* on the basis of "platitudes," imagine next that *we* are the culture being interpreted and that the interpreter is a member of a primitive culture. He could learn to understand English by coming to appreciate a large number of new facts, beliefs, ways of thinking. At the end of this, he could understand such an English sentence as "The sun is ninety-three million miles from the earth." But his understanding of this sentence would by no means be an application of platitudes that everyone in his tribe knows; for learning is not merely a matter of applying what one already knows to additional cases, but of making conceptual leaps, of projecting oneself imaginatively into new ways of thinking. One could never acquire a concept that the culture doesn't already have (or that isn't reducible to concepts a culture already has) if the platitude story were all there is to interpretation. Our platitudes are not a basis to which all possible notions can be reduced.

Lewis may now say that this belief isn't one that the primitive language can express, and that each culture only has to have a theory of the process of attributing beliefs that it can express. But this would be a bad reply, on two grounds. First, there are sentences that we can translate into the primitive language (I assume the primitive language has number words), for example, "If you were to travel the distance a fast horse can run in a day for ten thousand years, you still would not reach the sun," which we believe *because* we believe certain physical theories. The conditions under which we believe this sentence and the "motor responses" that belief in it causes us to exhibit are just inexplicable from a primitive point of view (without going beyond primitive platitudes); similarly, the conditions under which a sophisticated future culture might believe "There are intelligent beings in galaxies other than the Milky Way," or even "Snow is white," might be inexplicable from our point of view (without going beyond our platitudes). Second, as Lewis himself points out, a theory of the propositional attitudes must make sense of quantification over the propositional attitudes, not just of individual propositional attitudes that we are able to express in our current language.

That Lewis himself feels the force of something like these objections is indicated by the fact that in his one paper on interpretation[11] he does not claim that the propositional attitudes of another culture can be identified by relying on our platitudes, but instead introduces a list of a half-dozen "constraints" on interpretation, including principles of charity, rationality, and so on. These constraints are, however, extremely vague and themselves require a good deal of (so far unformalized) intuition to apply to any concrete case.

It may be that Lewis means to claim only that the propositional attitudes of speakers of our own language who share our culture can be identified by relying on our current platitudes; but would this establish his claim that the very *meaning* of "believes that snow is white" is "implicitly defined" by those platitudes? Alternatively, he may think that the "constraints" are themselves a loose summary of those of our platitudes that we use in assigning propositional attitudes to the speakers of another language. But if they are to *implicitly define* the propositional attitudes, there must be enough of these platitudes to fix the interpretation of the words that speakers of an arbitrary human language in an arbitrary human culture employ; and no reason has been given to think that there are platitudes which would enable us even to identify and translate the logical words in an arbitrary natural language. In a "regimented" language, the logical words can be singled out by the introduction and elimination rules that they obey (by such rules as "from p,q infer $p \mathbin{\&} q$," "from $p \mathbin{\&} q$ infer p," and "from $p \mathbin{\&} q$ infer q"). But the logical words of a natural language notoriously fail to satisfy such syntactically characterizable inference rules (because they also carry information about time order, causal connection, relevance, etc.).

In summary, the "platitudes" about belief, desire, action, that "everyone knows" do not amount to an implicit definition of the propositional attitudes, the desires, etc., for two reasons: the idea that there is one physical state per propositional attitude is false under any reasonably natural construal of "physical state," including Lewis's own "neurochemical state," except the wide construal used in the Appendix (if that counts as a "reasonably natural construal")— the construal under which there are so *many* physical states that "everything has every program"; and even if there *were* one physical state per propositional attitude, the platitudes of commonsense psychology are not enough to single out those physical states.

Conclusion

Functionalism started out by rejecting the naive idea that propositional-attitude descriptions (e.g., "believes that there are a lot of cats in the neighborhood") correspond one by one to brain states, species-independently. Yet—and this is its Achilles heel—it did assume that they correspond to brain states in this way *in each individual organism*.

But this is just a piece of "folk science"—in a pejorative sense. Viewed computationally, as devices for "confirming" or "accepting" sentences on the basis of sensory stimulation and for making "motor responses," human beings differ enormously from one another and from culture to culture. Although all human beings are computers of the same kind at the moment of birth, it is not the case that all adult human beings must go through the same sequence of states when they fix a belief that we would translate into our language by the sentence "There are a lot of cats in the neighborhood." Actual interpretative practice does not proceed by looking for something isolable, as "neurochemical states" are supposed to be isolable by their structure and biological and chemical roles independently of any semantics one might impose on them, but proceeds rather by *discounting differences.* Actual interpretative practice is open ended and practically infinitely extendable (to new cultures, new technologies, even new species, at least potentially). If one remembers that the only "handle" we actually have on the notion of "same belief" is interpretative practice, one will see that there is absolutely no reason to believe that there is one *computational* state that all possible human beings who think that "there are a lot of cats in the neighborhood" must be in. And if there is not a single *computational* state that they are all in, then there is not likely to be a relevant *neurochemical* state that they are all in either.

We have considered the reply that a being "smarter" than *Homo sapiens* might "in principle" survey all human modes of language construction, conceptualization, belief fixation, etc., and construct an equivalence class of computational states that any human being must be in to have a particular belief. But then the same problem arose again when we asked *what it is for this smarter being to have the belief.* (Or when we asked what "smarter" comes to here.) There is no reason to think that there is a *definable* equivalence relation over computational states that could provide equivalence classes corresponding to propositional attitudes (one equivalence class per propositional attitude) in the case of arbitrary physically possible intentional systems.

Finally, we considered the idea of letting the propositional attitudes be "implicitly defined" by theories. But we don't know what sort of theory could do this (even if the scope of the theory is restricted to *Homo sapiens*), nor do we have a notion of "implicit definition" that does not presuppose we know what *sort* of entity it is we wish to implicitly define. Thus Lewis and I independently took the course of assuming that there is one brain state per propositional attitude in the case of each individual organism.[12] But this is where we both fell victim to "folk science." So functionalism doesn't work. That is to say, it doesn't fit the phenomena. But much has been learned, I feel, by trying it on for size.

Chapter 7
A Sketch of an Alternative Picture

The end of an investigation like the one just completed is an appropriate time to reexamine the reasons for undertaking the investigation in the first place; for no philosophical investigation looks quite the same at the end as it did when one started. My reasons for undertaking this project were many. Some readers of the preceding pages have expressed surprise at what they called my "realist tone"; as though, by reacting to "mentalist" and "functionalist" proposals in the way that I have, I was renouncing my supposed "antirealism." Others have asked me to explain where the present inquiry "fits in" to my ongoing project of developing a third way ("internal realism") between classical realism and antirealism. In this closing chapter I want to respond to these requests and reactions.

Let me begin by admitting that I have long felt an approach/avoidance conflict where "metaphysical realism" is concerned. In various places I have described metaphysical realism as a bundle of intimately associated philosophical ideas about truth: the ideas that truth is a matter of Correspondence and that it exhibits Independence (of what humans do or could find out), Bivalence, and Uniqueness (there cannot be more than one complete and true description of Reality); but I don't think that this characterization caught the appeal of metaphysical realism to *me*—which was, of course, a grave defect. What I used to find seductive about metaphysical realism is the idea that *the way to solve philosophical problems is to construct a better scientific picture of the world.* That idea retains the ancient principle that Being is prior to Knowledge, while giving it a distinctively modern twist: all the philosopher has to do, in essence, is be a good "futurist"—anticipate for us *how* science will solve our philosophical problems. From this idea I was led naturally to the thought that science should be understood "without philosophical reinterpretation." In such an outlook, Independence, Uniqueness, Bivalence, and Correspondence are regulative ideas that the final scientific image is expected to live up to, as well as metaphysical assumptions that guarantee that such a final scientific resolution of all philosophical problems *must* be possible.

To fully understand the appeal of scientific realism, one does not necessarily have to know anything about science—Bernard Williams's *Ethics and the Limits of Philosophy*, for example, exhibits an enthusiasm for scientific realism ("the absolute conception of the world") coupled with complete innocence of actual scientific knowledge. But for someone like me who has mastered a certain amount of mathematics and physics, the situation is quite different. I *know* what a real scientific theory looks like; and the attractiveness of scientific realism is counterbalanced by an unwillingness to accept vague talk about what science can achieve as a substitute for at least a plausible sketch of a genuine scientific theory with real explanatory power. Hence what I described as my "approach/avoidance" conflict.

In working my way through this conflict, I found early on that the question of "intentionality" held a central position. That "physicalist" accounts of the world are incomplete—in particular, that they do not account for intentionality—is, to be sure, not a new claim. Brentano, Husserl, and others have made this claim, and Kant's remark[1] (in the *Critique of Pure Reason*) that we are unlikely to be able to give an account of "schematism" in natural scientific terms is itself an early version of this very claim. But few philosophers today[2] would expect a scientific realist account of intentionality to have the form that Kant, or Brentano, or Husserl might have been thinking of (that is, to be in terms of the "mechanics" of the brain or in terms of an associationist psychology of "ideas"). With the rise of computer science, an entirely new paradigm of what a scientific realist account of intentionality might look like presented itself. The need for a full-length investigation of the question of the scientific reducibility of intentionality in the age of the computer thus arose.

In the foregoing chapters I have tried to present a concise but adequately detailed account of such an investigation. Since my aim was to investigate the possibility of a scientific realist account in its *own* terms, rather than to criticize scientific realism (as a metaphysical position) from the standpoint of my own present position, I have tried to make my criticisms intelligible and just from the standpoint of someone who feels the appeal of scientific realism. But (in response to the requests I mentioned) I will close by indicating briefly what my own "positive" ideas are about the circle of problems we have been examining. It must be understood, however, that my purpose here will not be to engage in philosophical polemic in favor of the kind of pragmatic realism ("internal realism") I have been advocating, or to provide extended arguments for that position,[3] but simply to provide a minimum of information as to what my outlook is and how it bears on these issues.

Objectivity and Conceptual Relativity

What I want to present is not, indeed, a "theory" in the sense in which the scientific realist hopes that it will be possible to construct a theory of intentionality. I do not see any possibility of a scientific theory of the "nature" of the intentional realm, and the very assumption that such a theory *must* be possible if there is anything "to" intentional phenomena at all is one that I regard as wholly wrong. (My colleague Burton Dreben has long taught Harvard students of philosophy that it is just these philosophical "musts," just the points at which a philosopher feels that no argument is needed because something is just "obvious," that they should learn to challenge.) But I have not been quite persuaded by my Wittgensteinian colleagues that one should give up the effort to *explain* in philosophy. What I can offer is less than a theory, to be sure, but it is a picture that enables us to make some sort of sense of a variety of different phenomena. (Even Wittgenstein allowed himself to hope for a "perspicuous representation," after all.) The important thing, it seems to me, is to find a picture that enables us to make sense of the phenomena from within our world and our practice, rather than to seek a God's-Eye View.

The phenomena that have to be accounted for are partly familiar and partly unfamiliar (or long overlooked or unappreciated). So before I begin my sketch, I have to indicate what sorts of phenomena an adequate picture must enable us to make sense of. Many of those phenomena have been listed in the preceding chapters (division of linguistic labor, contribution of the environment, meaning holism, for example). Here I want to briefly discuss two very general desiderata for a philosophical picture of the intentional: it should account for both objectivity and conceptual relativity.

What do I mean by "objectivity and conceptual relativity"? To begin with objectivity: to say that intentional phenomena are "objective" is not to say that they are independent of what human beings know or could find out (it is not to say that they are Objective with a capital "O," so to speak). If we take "truth" as our representative intentional notion, then to say that truth is objective (with a small "o") is just to say that it is a property of truth that whether a sentence is true is logically independent of whether a majority of the members of the culture *believe* it to be true. And this is not a solution to the grand metaphysical question of Realism or Idealism, but simply a feature of our notion of truth.[4]

To be sure, this is a feature that has been challenged by cultural relativist philosophers. For example, although Rorty has since repented of this formulation,[5] in his *Philosophy and the Mirror of Nature* he defined truth in terms of the agreement of one's "cultural peers." But this feature of our notion of truth (and also of our notion of warrant) is one that cultural relativists themselves rely on, one that they themselves cannot help relying on in their practice. For the relativist, after all, knows perfectly well that the majority of *his* cultural peers do not accept his relativist views. But he does not conclude that his views must therefore be false, because he feels (perhaps unconsciously) that that is *irrelevant* to the question of the truth (and to the question of the warrant) of those views. I shall not press this point, because this is not the place to present a detailed argument against relativism.[6]

An eliminationist like Paul Churchland, who is willing to look for "a successor concept to the notion of truth," can, indeed, reject the objectivity of the notion of truth along with the notion itself. But the task of showing that a successor concept can be provided, that *it* has the kind of objectivity the scientific realist regards as characteristic of science, and that it can play a suitable role in explaining the success of scientific linguistic practice is today only a gleam in Churchland's eye.

Another way out is to account for the objectivity of the notion of truth (as well as of reference, etc.) in the way suggested by Brentano and by Chisholm; that is, just to take the existence of intentional properties as a "primitive" fact. If primitive just means "not reducible to nonintentional notions," then (I have been arguing) this is, indeed, the right answer. But if the idea is to take something like the traditional metaphysical realist notion of truth (and the metaphysical ideas of Correspondence, Independence, Bivalence, and Uniqueness associated with that notion of truth) as primitive, then my answer is "Not so fast!"

The reason I say "Not so fast!" is that there are *other* properties of truth, reference, and meaning to be accounted for than just the objective character of the notions. In particular there is what I call "conceptual relativity." Whereas objectivity, in one form or another, is not only a familiar property of the intentional notions, but the one most discussed by philosophers, conceptual relativity is a property which has only emerged as central in the twentieth century, and its very existence is still most often ignored, if not actually denied.

Let me begin with a very simple example. Suppose I take someone into a room with a chair, a table on which there are a lamp and a notebook and a ballpoint pen, and nothing else, and I ask, "How many objects are there in this room?" My companion answers, let us suppose, "Five." "What are they?" I ask. "A chair, a table, a lamp, a notebook, and a ballpoint pen." "How about you and me? Aren't we in the room?" My companion might chuckle. "I didn't think you meant I was to count people as objects. Alright, then, seven." "How about the pages of the notebook?"

At this point my companion is likely to become much less cooperative, to feel I have "pulled a fast one." But what is the answer to my question? A logician is likely to say that there is an ordinary (or perhaps a metaphysical) notion of an "object," according to which, perhaps, the pages of the notebook are not "objects" as long as they are still attached, and according to which my nose is not an object but only a part of an object as long as *it* is still attached (Aristotle would have said that a whole living person or animal was a "substance," but that a nose is only part of a substance, not a substance); and that there is a logical notion of an object or "entity" according to which anything we can take as a value of a variable of quantification (anything we can refer to with a pronoun)[7] is an "object"; and that all the parts of a person or a notebook are "objects" in this logical sense.

But even if we agree to use "object" in this "logical" sense, there turn out to be problems. Let us ignore quantum mechanics, and let us suppose there are exactly n elementary particles inside the room. Those n elementary particles are all "objects"; we can *refer* to them, or include them in the range of a variable of quantification. What about *groups* of elementary particles?

For the moment, let us take it that by a group we are to understand a whole with certain parts, not an abstract set (thus a group is what certain logicians[8] call a "mereological sum"). For example, my hand (we may suppose) is a group of atoms; those atoms are groups of elementary particles. What about the group consisting of my nose and the lamp? Is that an object at all? Or is there no such object?

Husserl, who first proposed the idea of a logical calculus of parts and wholes, thought that only certain "organic" wholes are real objects. My body is an object, but my nose is (phenomenologically) only a part of an object, not an object. But we are not asking what is phenomenologically an object, but what is an object in the logical sense. Is the mereological sum of my nose and the lamp an object in the logical sense?

If we say "No," then some philosopher will object that being "organic" is too subjective to serve as a criterion for what is and is not an object in the logical sense. "It is only our interests," the philosopher will object, "that make us regard the lamp as more than a strange discontinuous group of particles. To be sure, the parts of the lamp stay together when the lamp is moved (this was one of Aristotle's criteria for objecthood); but if a piece of chewing gum is stuck to a table, the sum of the chewing gum and the table also fulfills *that* criterion, and, moreover, some lamps do *not* stay together when moved. (The shade falls off.) Either you should consider only elementary particles to be objects, or you should allow arbitrary mereological sums."

If we agree that all mereological sums count as objects, we will say that there are 2^n "objects" in the room. If we count only "organic wholes" as objects, we will end up with a much smaller number. Which is right?

To me it seems clear that the question is one that calls for a convention. As a layman might well put it, "It depends on what you mean by an object." But the consequence is startling: the very meaning of existential quantification is left indeterminate as long as the notion of an "object in the logical sense" is left unspecified. So it looks as if *the logical connectives themselves have a variety of possible uses.*

The writ of convention runs farther than the decision to count/not to count mereological sums as objects, however. We have said that my nose is a group (mereological sum) of atoms. But Saul Kripke would deny this; he would say, "Since your nose could have consisted of different atoms, it has a modal property the group of atoms does not. So your nose is not identical with the group of atoms." David Lewis would reply that when we say that there is a possible world in which my nose consists of different atoms, what we mean is that there is a possible world in which a *counterpart* of my nose consists of different atoms. In *this* world, Lewis would say, my nose *is* identical with this group of atoms. Again it seems to me that the question calls for a convention. We can decide to speak with Kripke and we can decide to speak with Lewis and we can decide to speak in a variety of other ways (including deciding to say, "There is no fact of the matter as to whether the relation between the nose and the group of atoms is 'identity' or not").

Hans Reichenbach would have agreed that these questions call for conventions, and he would have added that it is precisely the job of the philosopher to distinguish what is fact and what is convention in our theorizing about the world.[9] But, as Quine later pointed out,[10] the very distinction between "fact" and "convention" on which Reichenbach relied collapses when construed as a sharp dichotomy. An example (my own, not Quine's) is the conventional character of any answer to the question "Is a point identical with a series of spheres that converges to it?" It is known since *Principia Mathematica* at least that we can identify points with sets of convergent spheres and all geometric facts will be correctly represented. We know that we can also take points as primitive and identify spheres with sets of points. So any answer to this question is, once again, conventional, in the sense that one is free to do either. But what Quine pointed out (as applied to this case) is that when I say, "We can do either," I am assuming a diffuse background of empirical facts. Fundamental changes in the way we do physical geometry could alter the whole picture. The fact that a truth is toward the "conventional" end of the convention–fact continuum does not mean that it is *absolutely* conventional—a truth by stipulation, free of every element of fact. And, on the other hand, even when we see such a "reality" as a tree, the possibility of that perception is dependent on a whole conceptual scheme in place (one which may or may not legislate an answer to such questions as "Is the tree identical with the space-time region that contains it?" and "Is the tree identical with the mereological sum of the time-slices of elementary particles that make it up?"). What is factual and what is conventional is a matter of degree. We cannot say, "These and these elements of the world are the raw facts, the rest is the result of convention."

Internal Realism as an Alternative Picture

I want to suggest a way of understanding precisely these sorts of facts. My suggestion, like every philosophical suggestion, has many forerunners: Carnap's important observation that the rules of formal logic do not uniquely determine the interpretation of the logical connectives[11] and Wittgenstein's aphorism "Meaning is use" are among them. The little example of my friend who says, "There are five objects in this room: the chair, the table, the lamp, the notebook, and the ballpoint pen," can serve to illustrate the idea.

There is a commonsense way of clearing up the puzzle about how many objects there are in the room, and that is to say, "It depends upon what you mean by 'object.'" This commonsense remark is perfectly right, but deeper than may appear to the commonsense mind itself.

As we saw, there are many ways of *using* the notion of an object— even the so-called "logical notion" of an object—or the existential quantifier. And, depending on how we use the notion, the answer to the question "How many objects are there in the room?" can be "Five," "Seven," "2^n"—and there are many more possibilities.

A metaphor which is often employed to explain this is the metaphor of the cookie cutter.[12] The things independent of all conceptual choices are the dough; our conceptual contribution is the shape of the cookie cutter. Unfortunately, this metaphor is of no real assistance in understanding the phenomenon of conceptual relativity. Take it seriously, and you are at once forced to answer the question "What are the various parts of the dough?" If you answer that (in the present case) the "atoms" of the dough are the n elementary particles and the other parts are the mereological sums containing more than one "atom," then you have simply adopted one particular transcendental metaphysical picture: the picture according to which mereological sums "really exist." My view—which I called "internal realism" in *Reason, Truth and History* (I would have done better to call it simply *pragmatic* realism)—denies that this is *more* the "right" way to view the situation than is—insisting that only the n elementary particles (or only the elementary particles and the atoms and molecules, or only the "organic wholes") really exist. The metaphysician who takes the latter view can also explain the success of the language of "mereological sums," after all: he can say that talk of mereological sums is really just a *façon de parler* (it is easy to "translate" such talk into set-theoretic talk, or into second-order quantification, etc., he could point out).[13]

The cookie-cutter metaphor *denies* (rather than explains) the phenomenon of conceptual relativity. The internal realist suggestion is quite different. The suggestion, applied to this very elementary example, is that what is (by commonsense standards) the same situation can be described in many different ways, depending on how we use the words. The situation does not itself legislate how words like "object," "entity," and "exist" must be used. What is wrong with the notion of objects existing "independently" of conceptual schemes is that there are no standards for the use of even the logical notions apart from conceptual choices. What the cookie-cutter metaphor tries to preserve is the naive idea that at least one Category—the ancient category of Object or Substance—has an absolute interpretation. The alternative to this idea is not the view that it's all *just* language. We can and should insist that some facts are there to be discovered and not legislated by us. But this is something to be said when one has adopted a way of speaking, a language, a "conceptual scheme." To talk of "facts" without specifying the language to be used is to talk of nothing; the word "fact" no more has its use fixed by the world itself than does the word "exist" or the word "object."

The seemingly more complex cases of conceptual relativity described above—the relativity of identity (as in the question "Is the tree identical with the space-time region it occupies?" or "Is the chair identical with the mereological sum of the elementary particles that make it up?") and the relativity of the categories Concrete and Abstract (as in the question "Is a space-time point a concrete individual, or is it a mere limit, and hence an abstract entity of some kind?")—and one might add many other examples—can all be handled in much the same way. "Identical," "individual," and "abstract" are notions with a variety of different uses. The difference between, say, describing space-time in a language that takes points as individuals and describing space-time in a language that takes points as mere limits is a difference in the choice of a language, and neither language is the "one true description."

The suggestion I am making, in short, is that *a statement is true of a situation just in case it would be correct to use the words of which the statement consists in that way in describing the situation.* Provided the concepts in question are not themselves ones which we ought to reject for one reason or another, we can explain what "correct to use the words of which the statement consists in that way" means by saying that it means nothing more nor less than that a sufficiently well placed speaker who used the words in that way would be fully warranted in *counting* the statement as true of that situation.[14]

What is "a sufficiently well placed speaker"? That depends on the statement one is dealing with. There is no algorithm for determining whether a given epistemic position is better or worse for making an arbitrary judgment. But facts of the form "If you have to tell whether S is true, then it is better to be in circumstances C_1 than in circumstances C_2" are not "transcendent" facts; they are facts that it is within the capacity of speakers to determine, if they have the good fortune to be in the right sorts of circumstances. What are "the right sorts of circumstances"? That depends on the statement one is dealing with. . . .

I am not being "cute." The point is that I am not offering a *reductive* account of truth, in any sense (nor of warrant, for that matter). In *Reason, Truth and History* I explained the idea thus: "truth is idealized rational acceptability." This formulation was taken by many as meaning that "rational acceptability" (and the notion of "better and worse epistemic situation," which I also employed) is supposed (by me) to be *more basic* than "truth"; that I was offering a *reduction* of truth *to* epistemic notions. Nothing was farther from my intention. The suggestion is simply that truth and rational acceptability are *interdependent* notions. Unfortunately, in *Reason, Truth and History* I gave examples of only one side of the interdependence: examples of the way truth depends on rational acceptability. But it seems clear to me that the dependence goes both ways: whether an epistemic situation is any good or not typically depends on whether many different statements are *true*.

To repeat: the suggestion which constitutes the essence of "internal realism" is that truth does not transcend use. Different statements— in some cases, even statements that are "incompatible" from the standpoint of classical logic and classical semantics—can be true in the same situation because the words—in some cases, the logical words themselves—are used differently. But this is not to say that talking of "use" instead of "meaning" is going to provide another sort of reductive account of or substitute for "intentionality." Describing the use of words involves describing many things—when sentences containing those words are acceptable, what typically causes an expert/ordinary speaker to use those words in particular ways, what interests the ways of speaking in question subserve, what the "phenomenology" of the particular way of speaking is, and so on. These things are no more reducible to physical-cum-computational language than is meaning talk or reference talk.

My Present Diagnosis of the "Functionalism" Issue

In closing, I want to say what the fundamental difficulties with scientific realist attempts to account for intentionality (attempts like my own "functionalism") really are. This diagnosis, as I have already indicated, will go beyond the arguments which constitute the body of this book; those arguments are meant as arguments from within the scientific realist's own perspective, arguments which the scientific realist must regard as presenting genuine difficulties from his own point of view. My own view of the real nature of the difficulties is from a different perspective. That does not mean that it depends through and through on the "internal realist" perspective just sketched. Part of what I will now say is still independent of that perspective. But the summing up with which I shall close certainly presupposes it.

First of all, let us notice that the difficulties with the "mentalist," "functionalist," physicalist-cum-functionalist, "sociofunctionalist," etc., programs just reviewed are of two kinds, or at least they are of two kinds from a scientific realist perspective. For a scientific realist, there is a gulf between epistemological and ontological issues, and scientific realism faces difficulties of both kinds.

The way in which I see the epistemological difficulties can be brought out with the aid of the following analogy. Consider the ordinary notion of a mathematical proof. This is not at all the same as the notion of proof in a formal system (any statement at all can be "proved" in a formal system, by just taking the statement itself, or any statements from which it follows by the rules of the system, as axioms). Nor is it the same as the notion of proof in a system which is sound (one whose axioms are true and whose rules of inference preserve truth). For if Fermat's Last Theorem is in fact true, then a system which has that statement (i.e., the statement "There do not exist positive integers $x, y, z, n > 2$, such that $x^n + y^n = z^n$") as its one and only axiom, and any truth-preserving rules of inference you like, is certainly a sound system in which Fermat's Last Theorem can be proved; but this is not what mathematicians call a "proof of Fermat's Last Theorem." A proof in the ordinary sense (a proof humanly speaking) is a proof in a system which is not just sound, but which a mathematician could, upon reflection, *see* to be sound, one which a reasonable mathematician would be justified in accepting. "Proof" is an *epistemic* notion, not a mathematical one.

Can this notion of proof itself be formalized? The question is delicate, but the answer practically all logicians would accept is the following: if there exists a sound system which does formalize this notion of proof (in the sense that each proof in that system is a proof in the ordinary sense, and all proofs in the ordinary sense can be reconstructed in that system)—and many logicians, like many philosophers, feel uncomfortable with this Platonic way of using "exists"—then *that* system *as a whole* is *not* one that a mathematician could, upon reflection, *see* to be sound!

The argument goes as follows: If a system S can, upon reflection, be seen to be sound, then one can prove the consistency of that system in a way which is also intuitively acceptable by arguing, "The axioms of S are true, and the rules of inference of S preserve truth; so all theorems of S are true; but '1 = 0' (or any formula of S that is obviously contradictory) is not true; so '1 = 0' is not a theorem of S; so S is consistent." (If one is willing to accept a language which is set-theoretically "stronger" than S whenever one is willing to accept S, then the notion of "truth" used in this argument can be replaced by its definition à la Tarski.) So whenever a system S is such that we can *see* that it is sound, then a "stronger" system S' (one in which we can carry out this little argument, either in its semantical version or in its set-theoretic version) is also such that we can *see* that it is sound. (This is often called a "Reflection Principle.") It follows that if system S is such that we can see that it is sound, then there exists a proof of the statement "S is consistent" (in the epistemic sense of "proof"). But if the proof of the statement "S is consistent" is a proof (in the epistemic sense), then so is a certain number-theoretic version of the proof (a proof of the Gödelian statement "CON S"). Since S was assumed such that *every* mathematical statement that has a proof has a proof in S, it follows that (if S can be seen to be sound by a human mathematician, then) "CON S" must be a theorem of S. But then, according to Gödel's Second Incompleteness Theorem, S must be *inconsistent*, contradicting the assumption that S is sound. Q.E.D.

While this falls short of a proof that no such system S "exists," it does show that *if* such a system exists—a system which computationally formalizes our intuitive notion of a mathematical proof—then this is a fact that we could not mathematically verify. This is not as devastating as it first looks, because it might be possible to verify that S has this property by an "empirical" argument: we might find that all the proofs we constructed in S were mathematically acceptable, and, after a time, we might accept S as a formalization of the intuitive notion of proof *without* claiming that it was, in any sense, "mathematically evident" that S was sound. In short, we might accept the statement that S has this property as a kind of "quasi-empirical hypothesis."

But now the question naturally arises, "Can similar arguments be constructed in inductive logic?" And the answer turns out to be yes.[15] Suppose S were a formal system of inductive logic, a formal theory of the relation "It is justified to degree r to believe P given evidence e"; then one can show that if our *entire* intuitive notion of "justification" is captured by S (so that if the hypothesis "S captures our intuitive notion of justification" is itself one that can be justified by empirical evidence, then the justificatory argument must itself in some way be formalizable within S), then the fact that this is so cannot be justified by *any* argument that an idealized human judge would be justified in accepting!

My purpose, in recalling these facts here, is *not* to suggest that one can give a formal ("Gödelian") argument to show that "functionalism doesn't work." The analogy I have in mind isn't a mathematical one. But notice what underlies these well-known Gödelian arguments. What Gödel showed is, so to speak, that we cannot fully formalize our own mathematical capacity *because it is part of that mathematical capacity itself that it can go beyond whatever it can formalize.* Similarly, my extension of Gödelian techniques to inductive logic showed that it is part of our notion of justification in general (not just of our notion of *mathematical* justification) that *reason can go beyond whatever reason can formalize.*

If we look at the arguments deployed against functionalism (and various other "isms") in the course of this work, we quickly see that they rest (or the "epistemological" arguments rest) on the same fact, though in a less formal way. The connection between the epistemological issues just mentioned and questions of reference and meaning is secured by the truth of meaning holism. As we saw in the first chapter of this work, reference is not just a matter of "causal connections"; it is a matter of *interpretation* (this was the point of the "phlogiston" example used in that chapter). And interpretation is an essentially holistic matter. A complete "formalization" of Interpretation, we argued, is as utopian a project as a complete "formalization" of Belief Fixation.

I can now restate this point in a somewhat different way: knowing what the words in a language mean (and without knowing what they mean, one cannot say what they *refer to*) is a matter of grasping the way they are *used*. But use is holistic; for knowing how words are used involves knowing how to fix beliefs containing those words, and belief fixation is holistic.

This does not mean that (as Dummett, Fodor, Derrida, and others have argued) if we accept meaning holism, then we must say that every time our procedures of belief fixation change, the meaning of every word in the language changes. (Dummett and Fodor regard this as a refutation of meaning holism; Derrida regards it as a consequence he accepts.) For "meaning is use" is not a *definition* of "meaning." Meanings are not objects in a museum, to which words somehow get attached; to say that two words have "the same meaning" (and/or "the same reference") is just to say that it is good interpretative practice to equate their meanings (or their reference). But sophisticated interpretative practice presupposes a sophisticated understanding of the way words are used by the community whose words one is interpreting.

In sum, the attempt to survey "meaning" or "reference" fails for the same reason the attempt to survey reason itself fails: reason can transcend whatever it can survey.

The epistemological argument just sketched does not mean that Mechanical Translation programs, "natural-language processing programs," and the like are impossible. Just as there can be more and less powerful formal systems of mathematics and more and less powerful formal systems of inductive logic, there can also be more and less powerful programs for interpreting utterances in a natural language. What this argument does suggest (if the analogy upon which it is based indeed holds) is that, just as no formal system of mathematics can *define* what it is to be a mathematical proof, and no formal system of inductive logic can define what it is to be "confirmed," so no program for interpreting utterances in a natural language can define what it is for utterances to be synonymous or even coreferential. A complete computational *characterization* of "proof," "confirmation," "synonymy," and so on, will always be an impossibility.

I said earlier that there are "ontological" as well as epistemological difficulties with the various scientific realist programs. These difficulties stem from the very nature of these programs. In describing one of the programs in question (sociofunctionalism) in chapter 5, I myself wrote, "Why not think of the entire society of organisms together with an appropriate part of its physical environment as analogous to a computer, and seek to describe functional relations within this larger system? Why not seek to characterize reference, in particular, as a functional relation between representations used by organisms and things which may be either inside or outside those organisms?" Although the discussion which followed focused on what I have just described as the "epistemological" difficulties, there is a remarkable "ontological" presupposition contained in the very statement of the project. *The project simply assumes from the outset that there is a single system ("the organisms and their physical environment") which contains all the objects that anyone could refer to.* The picture is that there is a certain domain of entities such that all ways of using words referentially are just different ways of singling out one or more of those entities. In short, the picture is that what an "object" of reference is is fixed once and for all at the start, and that the totality of objects in some scientific theory or other will turn out to coincide with the totality of All The Objects There Are.

But, from my "internal realist" perspective at least, there is no such totality as All The Objects There Are, inside or outside science. "Object" itself has many uses, and as we creatively invent new uses of words, we find that we can speak of "objects" that were not "values of any variable" in any language we previously spoke. (The invention of "set theory" by Cantor is a good example of this.) What looked like an innocent formulation of the problem—"Here are the objects to be referred to. Here are the speakers using words. How can we describe the relation between the speakers and the objects?"—becomes far from innocent when what is wanted is not a "natural-language processor" that works in some restricted context, but a "theory of reference." From an internal realist point of view, the very problem is nonsensical.

Of course, from my point of view the "epistemological" and the "ontological" are intimately related. Truth and reference are intimately connected with epistemic notions; the open texture of the notion of an object, the open texture of the notion of reference, the open texture of the notion of meaning, and the open texture of reason itself are all interconnected. It is from these interconnections that serious philosophical work on these notions must proceed.

Appendix

Theorem. Every ordinary open system is a realization of every abstract finite automaton.

Physical Principles. The proof I shall give requires the following two physical principles (which hold in classical physics when (1) the fields have no sources except particles; and (2) the number of point particles is at most denumerably infinite):

Principle of Continuity. The electromagnetic and gravitational fields are continuous, except possibly at a finite or denumerably infinite set of points. (Since we assume that the only sources of fields are particles, and that there are singularities only at point particles, this has the status of a physical law.)

Principle of Noncyclical Behavior. The system S is in different maximal states at different times. This principle will hold true of all systems that can "see" (are not shielded from electromagnetic and gravitational signals from) a clock. Since there are natural clocks from which no ordinary open system is shielded, all such systems satisfy this principle. (N.B.: It is not assumed that *this* principle has the status of a physical law; it is simply assumed that it is in fact true of all ordinary macroscopic open systems.)

In the sequel, we shall make use of the fact that this principle holds true both on the boundary of any ordinary open system (i.e., the state of the boundary of such a system is not the same at two different times) and a little way *inside* the boundary as well.

Lemma. If we form a system S' with the same spatial boundaries as S by stipulating that the conditions *inside* the boundary are to be the conditions that obtained inside S at time t while the conditions *on* the boundary are to be the ones that obtained on the boundary of S at time t', where $t \neq t'$ [note that this will be possible only if the spatial boundary assigned to the system S is the same at t and at t'], then the resulting system will violate the Principle of Continuity.

Proof (of the lemma): Every ordinary open system is exposed to signals from many clocks C (say, from the solar system, or from things which contain atoms undergoing radioactive decay, or from the system itself if it contains such radioactive material—in which latter case the system S itself coincides with the clock C). In fact, according to physics, there are signals from C from which it is not possible to shield S (for example, gravitational signals). These signals from C may be thought of, without loss of generality, as forming an "image" of C on the surface of S. For the same reason, there are also "images" of C *inside* the boundary of S. The "image" of C at, say, $t' = 12$ may be thought of as showing a "hand at the 12 position"; while the "image" of C at, say, $t = 11$ shows a "hand at the 11 position." Thus, for these values of t and t', the system S' would have a "12 image" on its boundary and an "11 image" at an arbitrary small distance inside its boundary; but this is to say that the fields which constitute the "images" would have a discontinuity along an entire continuous area, and hence at nondenumerably many points.

Proof of the Theorem. (I have stated the theorem in terms of finite automata, but the technique is easily adapted to other formalisms.) A finite automaton is characterized by a table which specifies the states and the required state-transitions. Without loss of generality, let us suppose the table calls for the automaton to go through the following sequence of states in the interval (in terms of "machine time") that we wish to simulate in real time: *ABABABA*. Let us suppose we are given a physical system S whose spatial boundary we have exactly defined, at least during the real-time interval we are interested in (say, a given 7-minute interval, e.g., from 12:00 to 12:07). We wish to find physical states A and B such that during the time interval we are interested in the system S "obeys" this table by going through the sequence of states *ABABABA*, and such that given just the laws of physics (including the Principle of Continuity) and the boundary conditions of S, a Laplacian supermind could predict the next state of the system (e.g., that S will be in state B from 12:03 to 12:04) given the previous state (given that S was in state A from 12:02 to 12:03). This will show that S "realizes" the given table during the interval specified. Since the technique of proof applies to *any* such table, we will have proved that S can be ascribed any machine table at all, and the description will be a "correct" one, in the sense that there really are physical states with respect to which S is a realization of the table ascribed.

I shall use the symbolic expression $St\,(S, t)$ to denote the maximal state of S at t (in classical physics this would be the value of all the field parameters at all the points inside the boundary of S at t). Let the beginnings of the intervals during which S is to be in one of its stages A or B be $t_1, t_2, \ldots, t_n$ (in the example given, $n = 7$, and the times in question are $t_1 = 12{:}00$, $t_2 = 12{:}01$, $t_3 = 12{:}02$, $t_4 = 12{:}03$, $t_5 = 12{:}04$, $t_6 = 12{:}05$, $t_7 = 12{:}06$). The end of the real-time interval during which we wish S to "obey" this table we call t_{n+1} ($= t_8 = 12{:}07$, in our example). For each of the intervals t_i to t_{i+1}, $i = 1,2, \ldots, n$, define a (nonmaximal) *interval state* s_i which is the "region" in phase space consisting of all the maximal states St (S, t) with $t_i \le t < t_{i+1}$. (I.e., S is in s_i just in case S is in one of the maximal states in this "region.") Note that the system S is in s_1 from t_1 to t_2, in s_2 from t_2 to $t_3, \ldots$, in s_n from t_n to t_{n+1}. (Left endpoint included in all cases but not the right—this is a convention to ensure the "machine" is in exactly one of the s_i at a given time.) The disjointness of the states s_i is guaranteed by the Principle of Noncyclical Behavior.

Define $A = s_1 \vee s_3 \vee s_5 \vee s_7$; $B = s_2 \vee s_4 \vee s_6$.

Then, as is easily checked, S is in state A from t_1 to t_2, from t_3 to t_4, from t_5 to t_6, and from t_7 to t_8, and in state B at all other times between t_1 and t_8. So S "has" the table we specified, with the states A,B we just defined as the "realizations" of the states A,B described by the table.

To show that being in state A at times t with $t_1 \le t < t_2$ "caused" S to go into state B during the interval $t_2 \le t < t_3$ (and similarly for the other state transitions called for by the table), we argue as follows: Given that S is in state A at a time t ($t_1 \le t < t_2$), and letting the maximal state of the boundary of S at that time t be B_t, it follows from the lemma that St (S, t) is the only maximal state in any of the "regions" (nonmaximal states) $s_1, s_2, \ldots, s_7$ that a system S under the boundary condition B_t could be in without violating the Principle of Continuity. (If the shape, size, or location of S changes with time, then unless S resumes the boundary it had at t at least once, the boundary of S at t will be the only boundary associated with any maximal state in the union of these regions which fits the boundary condition B_t, and the lemma is unnecessary.) *A fortiori*, St (S, t) is the only maximal state in A compatible with B_t. Hence, given the information that the system was in state A at t, and given the information that the boundary condition at t was B_t, a mathematically omniscient being can determine from the Principle of Continuity that the system S must have been in St (S, t), and can further determine, given the boundary conditions at subsequent times and the other laws of nature, how S evolves in the whole time interval under consideration. Q.E.D.

Discussion. When we model cognitive functions, we do not, of course, model them by means of automata without inputs and outputs.

Rather, we imagine that the "automaton" is connected with input devices—sensors, such as eyes or ears (or, in the simplest case, a "paper tape" on which the operator can print messages in a specified alphabet); and also connected with output devices—motor organs, speech organs, etc. (or, in the case originally imagined by Turing, another "paper tape" on which the automaton can print messages in another specified alphabet). These inputs and outputs have specified realizations, or at least their realizations must be of certain constrained kinds depending on our purposes; usually we are not allowed to simply *pick* physical states to serve as their "realizations," as we are allowed to do with the so-called "logical states" of the automaton.

If a physical object does not have motor organs or sensors of the specified kind, then, of course, it cannot be a model of a description which refers to a kind of automaton which, *ex hypothesi*, possesses motor organs and sensors of that kind. And even if it does possess such "inputs" and "outputs," it may behave in a way which violates predictions which follow from the description (e.g., print two "1"s in a row when it is a theorem that the machine with the given description never does this). So there is no hope that the theorem just proved will also hold, unchanged, for automata which have inputs and outputs which have been specified (or at least constrained) in physical terms.

Imagine, however, that an object S which takes strings of "1"s as inputs and prints such strings as outputs behaves from 12:00 to 12:07 exactly as *if* it had a certain description D. That is, S receives a certain string, say "111111," at 12:00 and prints a certain string, say "11," at 12:07, and there "exists" (mathematically speaking) a machine with description D which does this (by being in the appropriate state at each of the specified intervals, say 12:00 to 12:01, 12:01 to 12:02, . . . , and printing or erasing what it is supposed to print or erase when it is in a given state and scanning a given symbol). In this case, S too can be *interpreted* as being in these same logical states A,B,C, . . . at the very same times and following the very same transition rules; that is to say, we can find *physical* states A,B,C, . . . which S possesses at the appropriate times and which stand in the appropriate causal relations to one another and to the inputs and the outputs. The method of proof is exactly the same as in the theorem just proved (the unconstrained case). Thus we obtain that *the assumption that something is a "realization" of a given automaton description (possesses a specified "functional organization") is equivalent to the statement that it behaves as if it had that description.* In short, "functionalism," if it were correct, would imply behaviorism! If it is true that to possess given mental states is simply to possess a certain "functional organization," then it is also true that to possess given mental states is simply to possess certain behavior dispositions!

Notes

Introduction

1. "Meaning and Our Mental Life," in *The Kaleidoscope of Science*, ed. Edna Ullman-Margalit (Dordrecht: Reidel, 1986), 17–32; and "Meaning Holism," in *The Philosophy of W. V. Quine*, vol. 18 in *The Library of Living Philosophers*, ed. Lewis Hahn and Paul Schilpp (La Salle, Ill.: Open Court, 1986), 405–426.
2. Cf. my "Reflexive Reflections," *Erkenntnis* 22 (1985): 143–153.

Chapter 1

1. Lloyd Carr pointed out the inaccuracy in this attribution to me. Brentano did think that mental phenomena were characterized by being directed to "contents": thus topics 1, 3, and 4 in my list of topics included under the chapter heading "intentionality" are a central part of what Brentano was discussing. "Brentano's thesis" was meant by *him* to serve as a way of showing the autonomy of mentalistic psychology ("act-psychology") by showing that the mental was *separate* from the real (external) world. Brentano himself, to my knowledge, never used the word "intentionality," nor did he use the terms "intentional inexistence" and "intentional existence" to refer to the relation between mind and the real world, as philosophers have come to use the word "intentionality" after Husserl. It was Husserl, not Brentano, who saw in the intentionality of the mental that it provided a way of understanding how mind and world are related and how it is that in acts of consciousness we come to be directed to an *object*.
2. One of the few reliable discussions of this much-misinterpreted aspect of Wittgenstein's thought occurs in Rush Rhees's *Discussions with Wittgenstein* (New York: Schocken, 1970). See esp. 46ff.
3. *The Language of Thought* (New York: Thomas Y. Crowell, 1975).
4. See, for example, Chomsky's *Reflections on Language* (New York: Pantheon, 1975), chap. 1.
5. Chomsky speaks of "a subsystem [for language] which has a specific integrated character and which is in effect the genetic program for a specific organ" in the discussion with Piaget, Pappert, and others reprinted in *Language and Learning*, ed. Massimo Piatelli (Cambridge, Mass.: Harvard University Press, 1980). See also *Reflections on Language*.
6. Or "informationally encapsulated input-output devices."
7. See chap. 5 of my *Mind, Language and Reality*, vol. 2 of my *Philosophical Papers* (Cambridge: Cambridge University Press, 1975).
8. Review of Skinner's *Verbal Behavior, Language* 35 (1955): 26–58.
9. See his *Problems of Knowledge and Freedom* (New York: Pantheon, 1971).
10. What gripped me, of course, was the idea of merging belief-desire psychology and

computational psychology; I was never attracted to the Innateness Hypothesis as the way to do this.

11. Of course, the brain's innate "language," if it exists, is not literally *written*. See "What Is Innate and Why," chap. 14 of *Mind, Language and Reality*.

12. "Two Dogmas of Empiricism," originally published in the *Philosophical Review* in January 1951; reprinted in Quine's *From a Logical Point of View* (Cambridge, Mass.: Harvard University Press, 1961); "Carnap on Logical Truth," originally published in *The Philosophy of Rudolf Carnap* (La Salle, Ill.: Open Court, 1963); reprinted in Quine's *Ways of Paradox* (Cambridge, Mass.: Harvard University Press, 2d ed. 1976); *Word and Object* (Cambridge, Mass.: MIT Press, 1960).

13. See, for example, his *Psychological Explanation* (New York: Random House, 1968) and *The Modularity of Mind* (Cambridge, Mass.: MIT Press, 1983), esp. the last 45 pages.

14. *Reason, Truth and History* (Cambridge: Cambridge University Press, 1981); a number of the papers in *Realism and Reason*, vol. 3 of my *Philosophical Papers* (Cambridge: Cambridge University Press, 1983). See also my "Reflexive Reflections," *Erkenntnis* 22 (1985).

15. See chap. 13 of my *Mind, Language and Reality*.

16. I am, of course, referring here to the relativistic views of Thomas Kuhn, whose *The Structure of Scientific Revolutions* (2d ed., enlarged, Chicago: University of Chicago Press, 1983) has become a classic.

17. In the metatheory of ordinary "extensional" logic—propositional calculus, quantification theory, set theory—one does not, indeed, need the notion of sameness of *sense*, most authors say, but the notions of extensional semantics—denotation or reference—play a conspicuous role in all treatments. I shall argue in the next chapter that any theory of sameness of "sense" must presuppose a theory of sameness of reference.

18. An interesting defense against this objection has been put forward by Massimo Piattelli-Palmarini in "The Rise of Selective Theories: A Case Study and Some Lessons from Immunology," chap. 6 in *Language Learning and Concept Acquisition: Foundational Issues*, ed. William Demopoulos and Ausonio Marras (Norwood, N.J.: Ablex, 1986). Piattelli points out that when Niels Kai Jerne first proposed that the body produces antibodies in *all possible configurations* (in 1955), there was enormous reluctance to believe this. Yet this hypothesis has subsequently been verified. "Niels Jerne postulates the *actual* presence in each organism of an 'internal image' of *any possible molecular pattern*" (p. 127). "It is hard to find arguments that the case must be simpler in cognitive matters" (p. 129).

I am not, however, inclined to accept this intriguing suggestion. Suppose one writes a "short story" (or rather, a plan for a short story) in such a way that at each of 200 places there is the option of performing two different actions. Each of the stories generated by the outline might be only two or three pages long—short enough to be memorized by a human being. Yet the total number of short stories generated by the outline is much too large—larger than the number of elementary particles in the universe—for all of the short stories to be individually written out anywhere. There is an enormous difference in cardinality between the number of "contents" a human being can learn and the number of antibodies in the human bloodstream. (The number of antibodies in the human bloodstream is certainly less than 10^{17}.) To be sure, "concepts" are not short stories. But they often arise from *theories*, and (by a similar argument) the number of possible theories (and, I suspect, theory *types*)—even of theories that are relatively "short"—involves an exponential explosion that makes the idea that evolution exhausted all the possibilities in advance wildly implausible.

Chapter 2

1. In a variant of the picture—one that represents the, so to speak, legacy of Plato rather than that of Aristotle—"concepts" are not in the mind, but rather form a realm of abstract entities (sometimes called "Platonic heaven" by detractors of the picture) independent both of the mind and of the world. Such a Platonism was, for example, defended by the great logician Kurt Gödel. Even in these "Platonistic" versions, however, speakers are supposed to be able to direct their mental attention *to* concepts by means of something akin to perception, and, if A and B are different concepts, then attending to A and attending to B are different mental states. So even in these theories, the mental state of the speaker determines which concept he is attending to, and thereby determines what it is he refers to.

2. I first argued this at length in "The Meaning of 'Meaning,'" chap. 12 of *Mind, Language and Reality* (originally published in *Language, Mind and Knowledge*, ed. K. Gunderson, Minnesota Studies in the Philosophy of Science, 7 (Minneapolis: University of Minnesota Press, 1975).

3. Cf. his *Varieties of Reference* (Oxford: Oxford University Press, 1982).

4. Evans's view in *Varieties of Reference* (as I understand it), as applied to the case of "gold," would be that the "concept" of gold is simply the ability to single out gold. This is possessed by all the experts I described above, and it is the *same* ability (since it is the ability to pick out the *same* stuff). Although Evans calls himself a mentalist, this is not mentalism in the sense of Fodor and Chomsky, since concepts are individuated by *stuff-involving* and *object-involving* abilities, not by the "syntax" of representations *inside* the mind/brain. My own theory is like Evans's in holding that the concept is partly individuated by the *stuff in the world* or *objects in the world* it applies to; but I reject the view that one must be able to identify the stuff *oneself* to be said to have the concept.

5. Cf. "The Meaning of 'Meaning'" (cited in n. 2).

6. Cf. his *Intentionality* (Cambridge: Cambridge University Press, 1983).

7. Including a public discussion following the reading of a paper by myself (titled "Why Meanings Aren't in the Head") at Rutgers University on March 13, 1986.

8. Cf. his *Minds, Bodies and Science* (Cambridge, Mass.: Harvard University Press, 1984).

9. I do discuss these questions at length in my *Reason, Truth and History* (Cambridge: Cambridge University Press, 1981), and also in my *Realism and Reason*, vol. 3 of my *Philosophical Papers*.

10. To see vividly what this means, imagine that somehow in Nova Scotia the words "elm" and "beech" have gotten switched. Then, on Searle's theory, it is wrong to say that the word "elm" in American English has the same intension as the word "beech" in Nova Scotian English and the word "beech" in American English has the same intension as the word "elm" in Nova Scotian English. In less technical language, what Searle can't say is, "In Nova Scotia 'elm' means *beech* and 'beech' means *elm.*"

11. The case of the Thai word for cat ("meew") shows, on the other hand, that even the kind of mental representation we have considered—perceptual prototype—needn't be precisely preserved in translation. What we ask in translation, even when perceptual prototypes are relevant, is not that they be the same but that they be sufficiently similar.

12. "The Meaning of 'Meaning.'"

13. I am indebted here to Jaak van Brakel, although I have not been able to accept his own view that it is constancies in the "phenomenological" properties—e.g., the melting point and the freezing point—that fix the reference of substance terms.

14. Berger introduces the notion of "focusing" in *Terms and Truth* (Cambridge, Mass.: MIT Press, 1988).
15. Tyler Burge objected to my expressing this (in "The Meaning of 'Meaning'") by saying that natural-kind terms have an "indexical component," on the grounds that they are not *synonymous with descriptions containing indexical words*. But I never claimed that they were. (See his "Individualism and the Mental," in *Midwest Studies in Philosophy* 10 (1986), ed. P. French, T. Euhling, and H. Wettstein.
16. This point was first emphasized by David Wiggins. See Wiggins's *Sameness and Substance* (Oxford: Blackwell, 1980).
17. This account was developed independently of my own. My own was first presented in lectures at Harvard in 1967–1968, and at lectures at Seattle and at the University of Minnesota the following summer; neither Kripke nor I published our accounts until a few years later.
18. The reader interested in problems having to do with modal contexts, counterfactuals, etc., will observe that I have discussed only the question of the reference of these words in the *actual* world. I believe that a similar account can be given of their reference in all *physically possible* worlds; Kripke's claim that the account extends even farther, to the fixing of reference in what he calls "metaphysically possible worlds" (which may not obey the same laws of nature as the actual world), now appears problematic to me.
19. However, in a recent paper ["Banish DisContent," in *Language, Mind, and Logic*, ed. Jeremey Butterfield (Cambridge: Cambridge University Press, 1986), 1–23] Fodor has backslid on this point—in this paper he speaks of the *function* from context to referent (e.g., the function that assigns H_2O to the word "water" on Earth and XYZ to the word "water" on Twin Earth) as being in the speaker's "head," and *this*—the function—is now identified with the "narrow content." This resembles Searle's view, and is open to exactly the same objections.
20. I write "identity" here and not "equivalence" to simplify the exposition. Strictly speaking, what Fodor's theory requires is not that sentences with identical meaning have numerically the *same* underlying "semantic representation," but that there be some *syntactically definable* and computationally effective) *equivalence relation* that holds between two expressions in "Mentalese" when and only when they are synonymous.
21. See Fodor's *RePresentations* (Cambridge, Mass.: MIT Press, 1981).
22. The term "ambiguous" is misleading here, as a report of Fodor's view (which is why I write "*referentially* ambiguous"). The term might suggest that Fodor's is really a conceptual claim, and Fodor does not intend his work as a conceptual analysis, but as a scientific theory. The claim that "meaning" refers to two different things (if Fodor is right) should be understood as an empirical claim, not a conceptual one.

Chapter 3

1. As of this chapter's writing—September 1986—Fodor's latest view was the view in "Banish DisContent." Robert Stalnaker has laid out an incisive criticism of this view in a paper ("On What's in the Head") which is unpublished as yet. See also Fodor's *Psychosemantics* (Cambridge, Mass.: MIT Press, 1987).
2. See n. 19 to the previous chapter.
3. Although this paper was not published until 1987 (in Hahn and Schilpp, *The Philosophy of W. V. Quine*), it circulated among my friends for quite a few years before publication. This explains how Fodor was able to prepare a sequence of replies to a paper which was only published later.
4. See Fodor's *The Modularity of Mind*, esp. 94ff.

5. Cf. Ned Block, ed. *Imagery* (Cambridge, Mass.: MIT Press, 1981).

6. That "narrow content," defined this way, can contain even *this* information independently of the environment has been challenged by Ernest LePore and Barry Loewer. See "Solipsistic Semantics," in *Midwest Studies in Philosophy* 10 (1986), ed. P. French, T. Euhling, and H. Wettstein.

7. Or with the "function of the observable properties" that the module is designed to detect—this is the way Fodor himself proposed to describe narrow content in the period when he entertained this view.

8. Fodor's "Cognitive Science and the Twin-Earth Problem," *Notre Dame Journal of Formal Logic*, 23, no. 2 (April 1982): 98–118.

9. As I understand it, Fodor *denies* this—he thinks that there is one *universal* stereotype of a dog, of a cat, etc. But the evidence he cites for this in *The Modularity of Mind* is simply not relevant to this claim. On this, see my review in *Cognition* 17 (1984): 253–264.

10. Ned Block, "An Advertisement for a Semantics for Psychology," in *Midwest Studies in Philosophy* 10.

11. In particular, Gilbert Harman, Wilfrid Sellars, and Hartry Field.

12. See Sellars's *Science, Perception and Reality* (Atlantic Highlands, N.J.: Humanities Press, 1963).

13. Since they apply/fail to apply to the same objects in all possible worlds, terms that are logically equivalent have the same "broad content." On some theories, terms that are not logically equivalent but "metaphysically equivalent," e.g., "water" and "H_2O," are also said to have the same "broad content." Yet a speaker may believe that there is water in the ocean and not believe that there is H_2O in the ocean, or believe that he has two hands and not believe that the number of his hands is an even prime. This motivates Block's search for a notion of "narrow content"; narrow-content differences are the only "meaning differences" there can be between terms that have the same "broad meaning," in his scheme.

14. These words, it should be pointed out, not only lack a (nonempty) extension; they also lack any clear broad content, that is, we have no clear way of deciding what a "possible world" in which there really was phlogistion, or caloric, or ether, would look like.

15. Another difficulty for Block is that someone may regard a belief as stereotypical without actually *having* the belief. I don't really infer that present-day kings *rule* from the fact that they are kings; but I know that this is part of the stereotype. Yet conceptual role is defined in terms of the inferences we make.

Chapter 4

1. See, in particular, *Word and Object* (Cambridge, Mass.: MIT Press, 1960) and the papers collected in *From a Logical Point of View* (Cambridge, Mass.: Harvard University Press, 1953).

2. Stephen Stich, *From Folk Psychology to Cognitive Science: The Case Against Belief* (Cambridge, Mass.: MIT Press, 1983).

3. Paul Churchland, "Eliminative Materialism and Propositional Attitudes," *Journal of Philosophy* 78, no. 2 (1981); Patricia Churchland, *Neurophilosophy: Toward a Unified Science of the Mind-Brain* (Cambridge, Mass.: MIT Press, 1986).

4. I argue this in my Carus Lectures, *The Many Faces of Realism* (La Salle, Ill.: Open Court, 1987).

5. I describe my conversation with Paul Churchland in a discussion which is reprinted verbatim in Zenon Pylyshyn and William Demopoulos, eds., *Meaning and Cognitive*

Structure: Issues in the Computational Theory of Mind (Norwood, N.J.: Ablex, 1986). See esp. page 244 and Churchland's clarification of his position on page 252.

6. Alfred Tarski, "The Concept of Truth in Formalized Languages," in his *Logic, Semantics, Metamathematics: Papers from 1923 to 1938* (Oxford: Clarendon Press, 1956), 152–278.

7. In conversation, Quine has admitted that he finds this *very* counterintuitive indeed!

8. This is the form mentioned in Tarski's famous "Convention T."

9. See n. 5 to this chapter.

10. This suggestion is considered (but not finally accepted) by Hartry Field in a searching paper, "Deflationary Theories of Truth," in *Fact, Science and Morality*, ed. G. MacDonald and C. Wright (Oxford: Basil Blackwell, 1986), 55–116. Field ascribes the view to a number of different philosophers.

Chapter 5

1. These papers are reprinted as chaps. 14, 18, 19, 20, and 21 of *Mind, Language and Reality*.

2. I explain my rejection of Turing machines as a model for the mind in "Philosophy and Our Mental Life," chap. 14 of *Mind, Language and Reality*.

3. The idea of sociofunctionalism was advanced by Richard Boyd in a (so-far unpublished) lecture a few years ago.

4. Lewis's views are discussed in the next chapter.

5. The idea that the irreducibility of intentional idioms to nonintentional ones is analogous to the irreducibility of material-thing notions to sense-datum notions was advanced by Roderick Chisholm in a famous correspondence with Wilfrid Sellars many years ago. See the "Chisholm-Sellars Correspondence on Intentionality," in *Concepts, Theories, and the Mind-Body Problem*, ed. Herbert Feigl, Michael Scriven, and Grover Maxwell, Minnesota Studies in the Philosophy of Science, 2 (Minneapolis: University of Minnesota Press, 1958), 521–539.

6. I point out in "Reference and Understanding," part 3 of my *Meaning and the Moral Sciences* (London: Routledge and Kegan Paul, 1978), that such a model is contained in the work of Carnap and Reichenbach. For a further discussion of the model, see the paper cited in n. 7 to this chapter.

7. "Computational Psychology and Interpretation Theory," in *Realism and Reason*, vol. 3 of my *Philosophical Papers* (Cambridge: Cambridge University Press, 1983).

8. The *locus classicus* for the discussion of "grue" is Goodman's *Fact, Fiction and Forecast* (4th ed., Cambridge, Mass.: Harvard University Press, 1983).

9. Cf. *Judgment under Uncertainty:Heuristics and Biases*, ed. Daniel Kahneman, Paul Slovic, and Amos Twersky (Cambridge: Cambridge University Press, 1981).

10. The notion of a "trial and error predicate" was introduced in my "Trial and Error Predicates and the Solution to a Problem of Mostowski," *The Journal of Symbolic Logic* 30, no. 1 (March 1965): 49–57. Such predicates are limits of recursive predicate; their use is possible if one does not ask that one be able to *know* when one's estimate of the value of the predicate has converged, but only that it will sooner or later converge.

11. "Arithmetical relations" are the relations in the finite levels of the Kleene hierarchy; they are definable using quantifiers over natural numbers, but no quantifiers over sets of natural numbers. Trial and error predicates, recursive predicates, and recursively enumerable predicates are all "arithmetical" in this sense.

Chapter 6

1. Some of these are reprinted as part 2 of David Lewis, *Philosophical Papers*, vol. 1 (Oxford: Oxford University Press, 1983). "Psychophysical and Theoretical Identification," cited in n. 2 below, was published in *Australasian Journal of Philosophy* 50 (1972): 249–258.
2. "Psychophysical and Theoretical Identification," 256.
3. "An Argument for the Identity Theory," in Lewis's *Philosophical Papers*, vol. 1.
4. "Neural state" is a term Lewis uses both in "An Argument for the Identity Theory" and in "Psychophysical and Theoretical Identification"; in the latter paper the identity theory of "An Argument for the Identity Theory" is broadened from "experiences" to "mental states" in general.
5. "Psychophysical and Theoretical Identification," 257. (Compare "Mad Pain and Martian Pain" in *Philosophical Papers*, vol. 1.)
6. "Psychophysical and Theoretical Identification," 256.
7. Some of Lewis's causality papers are reprinted as part 6 of *Philosophical Papers*, vol. 2 (Oxford: Oxford University Press, 1986).
8. "New Work for a Theory of Universals," *Australasian Journal of Philosophy* 61 (1983): 343–377.
9. "Elite" sets are introduced in a paper by Lewis entitled "Putnam's Paradox," *Australasian Journal of Philosophy* 62 (1984): 221–236. The notion seems to be the same as the notion called "naturalness" in "New Work for a Theory of Universals."
10. Reprinted in vol. 2 of my *Philosophical Papers*.
11. "Radical Interpretation," in Lewis's *Philosophical Papers*, vol. 1.
12. Lewis makes this assumption quite explicit in "Mad Pain and Martian Pain." To assert that creatures have pain—and what goes for pain and other experiences also goes for "mental states in general," he asserts—is to assert that there is a state which plays a certain causal role in an "appropriate population" of those creatures; in the case of humans, the state is a "neural state."

Chapter 7

1. Kant in the first Critique—*Critique of Pure Reason*, A140–42/B179–81—makes the interesting remark that the schematism of our understanding in its application to appearances is "an art concealed in the depths of the human soul, whose real modes of activity nature is hardly likely ever to allow us to discover."
2. Searle is an exception; in *Minds, Bodies and Science* he asserts that intentionality will be explained in terms of the physical chemistry of the brain just as the liquidity of water has been explained in terms of its physical chemistry. Needless to say, he gives no details.
3. My arguments for "internal realism" are spelled out in *Reason, Truth and History; Realism and Reason;* and *The Many Faces of Realism*.
4. When I say this, I do not mean that it is *analytic* that truth and acceptability to the majority of one's cultural peers are independent properties; but I do mean that this is a central feature of our picture of truth. The fact that a philosophically useful analytic/synthetic dichotomy cannot be drawn (because, for one thing, most of the things a philosopher would say are "conceptual truths" have, in one way or another, empirical *presuppositions*) does not mean that the notion of conceptual truth must be totally abandoned. It means, rather, that conceptual truth is a matter of *degree*.
5. See his *Consequences of Pragmatism* (Minneapolis: University of Minnesota Press, 1982), esp. the preface (page xxv).

6. My arguments against relativism are set out in *Reason, Truth and History.*

7. The idea of comparing variables in a formalized language to pronouns in this way is, of course, due to Quine.

8. See, for example, Quine and Goodman's "Steps toward a Constructive Nominalism," *The Journal of Symbolic Logic* 12 (1946): 97–122. Perhaps the first logician to formalize the logic of mereological sums was Leśniewski.

9. Cf. Reichenbach's *Philosophy of Space and Time* (New York: Dover, 1958; originally published in German in 1924).

10. Quine's "Two Dogmas of Empiricism," reprinted in his *From a Logical Point of View.*

11. This claim is the thesis of Carnap's *The Formalization of Logic,* now in print as the second part of *Introduction to Semantics and Formalization of Logic* (Cambridge, Mass.: Harvard University Press, 1959).

12. The discussion of the cookie-cutter metaphor in this paragraph is taken from the second of my Carus Lectures (*The Many Faces of Realism*).

13. When the number of individuals is fixed at some finite number n, then the mereological-sums language and the language in which we quantify only over atoms can easily be seen to be intertranslatable. For a discussion of the significance of this see *The Many Faces of Realism.*

14. The proviso "provided the concepts are not ones which we ought to reject for one reason or another" is important here; it is precisely the mistake of cultural relativism, in many of its forms, to ignore the fact that rejecting the concepts which are current in a particular "culture" at a particular time can be a *reform* and not just a *change.* In saying this I am, of course, rejecting the "fact/value" dichotomy which underlies many versions of relativism.

15. See "Reflexive Reflections" for details. An earlier application of Gödelian techniques to inductive logic occurs in my "'Degree of Confirmation' and Inductive Logic," in *The Philosophy of Rudolf Carnap,* ed. P. A. Schilpp (La Salle, Ill.: Open Court, 1963). (Reprinted in my *Mathematics, Matter and Method,* vol. 1 of my *Philosophical Papers.*)

Author Index